Presented to

Clare College, Cambridge

by

E. J. ANDERSON

LINEAR PROGRAMMING IN INFINITE-DIMENSIONAL SPACES

WILEY-INTERSCIENCE
SERIES IN DISCRETE MATHEMATICS AND OPTIMIZATION

Graham, Rothschild, and Spencer
RAMSEY THEORY

Tucker
APPLIED COMBINATORICS

Pless
INTRODUCTION TO THE THEORY OF ERROR-CORRECTING CODES

Nemirovsky and Yudin
PROBLEM COMPLEXITY AND METHOD EFFICIENCY IN OPTIMIZATION
(Translated by E. R. Dawson)

Goulden and Jackson
COMBINATORIAL ENUMERATION

Gondran and Minoux
GRAPHS AND ALGORITHMS
(Translated by S. Vajda)

Fishburn
INTERVAL ORDERS AND INTERVAL GRAPHS

Tomescu
PROBLEMS IN COMBINATORICS AND GRAPH THEORY
(Translated by R. A. Melter)

Palmer
GRAPHICAL EVOLUTION

Lawler, Lenstra, Rinnooy Kan, and Shmoys
THE TRAVELING SALESMAN PROBLEM

Hall
COMBINATORIAL THEORY

Minoux
MATHEMATICAL PROGRAMMING

Schrijver
THEORY OF LINEAR AND INTEGER PROGRAMMING

Anderson and Nash
LINEAR PROGRAMMING IN INFINITE-DIMENSIONAL SPACES

LINEAR PROGRAMMING IN INFINITE-DIMENSIONAL SPACES

Theory and Applications

Edward J. Anderson
Peter Nash

Management Studies Group
Engineering Department, University of Cambridge

A Wiley–Interscience Publication

JOHN WILEY & SONS
Chichester · New York · Brisbane · Toronto · Singapore

Library of Congress Cataloging in Publication Data:

Anderson, E. J. (Edward J.), 1954–
Linear programming in infinite-dimensional spaces.

(Wiley–Interscience series in discrete mathematics and optimization)
'A Wiley–Interscience publication.'
Includes index.
1. Linear programming. 2. Vector spaces.
3. Duality theory (Mathematics) I. Nash, Peter.
II. Title. III. Series.
T57.74.A467 1987 519.7′2 86-32579

ISBN 0 471 91250 6

British Library Cataloguing in Publication Data:

Anderson, Edward J.
Linear programming in infinite-dimensional spaces. —(Wiley Interscience series in discrete mathematics and optimization)
1. Linear programming
I. Title II. Nash, Peter
519.7′2 T57.74

ISBN 0–471 91250 6

Printed and bound in Great Britain

Preface

A linear program is an optimization problem with linear objective function and linear constraints. Problems of this type are of central importance in the theory and practice of optimization, for several reasons. Principally, the theory of linear programming in finite-dimensional spaces is both elegant and complete, an appealing embodiment of Lagrangian duality theory. The simplex algorithm and all its variations are very efficient practical solution methods, and provide not just an optimal solution but very useful information about the sensitivity of the solution to variations in the problem data, crucial when these data are known only imprecisely. In consequence, it is often found worthwhile to make linear models of non-linear real-world problems, despite the loss of accuracy that this involves.

In recent years, there has been a movement towards the integration of a number of seemingly different areas of optimization theory through the study of optimization problems posed in an abstract setting, capable of encompassing both finite- and infinite-dimensional problems. A landmark in this area is the book *Optimization by Vector Space Methods*, by D. G. Luenberger, published in 1969. This book is a convincing demonstration of the power of an abstract approach, and the way in which it can deepen our understanding of optimization theory.

It is probably inevitable, then, that even without the impetus of applications, linear programming in an abstract setting would have been a subject of study. In fact, infinite-dimensional linear programs arise in a number of applications, and it was from these that the first attempts to extend the finite-dimensional theory sprang. These early attempts were more or less equally concerned with duality theory and with the development of extensions of the simplex algorithm. Latterly, work has tended to concentrate rather more on the former than the latter, although both remain active research areas.

At the time of writing of this book, a fairly extensive theory exists for so-called semi-infinite programming, which is concerned with problems in which either the number of variables or the number of constraints is infinite, but not both. This theory encompasses both duality and solution algorithms. In the wider setting of infinite-dimensional linear programming, no such complete theory exists, and the work on duality theory and that on simplex extensions have not so far been integrated to the same extent. The main objectives of this book are to survey the theory as it now stands, and to attempt as far as possible to present these two areas of work in a way which brings out the connections between them and their implications for each other. In doing this we treat a number of interesting problems arising from very disparate areas of mathematics.

We are particularly concerned to try to elucidate the way in which the properties of linear programs and the workings of simplex-like algorithms depend on the underlying properties of the spaces in which the program is set. Accordingly, the book begins with an examination of the sorts of infinite-dimensional problem that arise in applications, and examines what happens when we try to extend the concepts of finite linear programming to them in a naïve way. Numerous difficulties appear, and in Chapter 2 we attempt to resolve these by formulating appropriate definitions of the concepts of basis and degeneracy. In doing this, we postulate a bare minimum of structure in the spaces in which our programs are posed: linear-space structure, a partial order, a linear objective functional and a linear constraint map. These suffice for some sort of strong duality result for general problems, though not necessarily of practical value. More useful strong duality results are derived in Chapter 3, where we make assumptions about the topological structures of the problem spaces, and use functional analytic methods to establish the existence of optimal dual solutions.

Chapter 4 contains a brief treatment of semi-infinite programming. This by its nature lies half way between finite LP and the fully infinite problems discussed in the following chapters. In Chapters 5–7, three particular classes of infinite-dimensional linear programs are examined in some detail. For each, the duality theory is developed, illustrating the theory described in Chapters 2 and 3, and used in the formulation of solution algorithms. These classes of programs are infinite-dimensional analogues of the assignment problem, the maximal flow in a network problem, and a problem related to the minimum-cost multi-commodity network flow problem, which we have called the separated continuous linear program. This progression is an attempt to go from problems with a lot of structure, and work towards the more general. The final chapter of the book gives a very brief discussion of four other types of infinite-dimensional linear program.

Our interest in infinite-dimensional linear programs has grown out of work on algorithms for the solution of some special classes of infinite-dimensional linear programs. This has meant that our concern is with linear programming rather than just linear programs. However, we have chosen not to include descriptions of detailed numerical methods.

This book could be used as the text for a course of lectures on infinite-dimensional LP, or in conjunction with other books in a more general course on optimization in general vector spaces. An early draft was in fact used for the first purpose here in Cambridge. The prerequisites for the book are an exposure to optimization theory in R^n (and in particular, linear programming) and elementary convex analysis, linear algebra and functional analysis.

Much of this book, especially the sections dealing with algorithms, has come out of work done in Cambridge over the last four years. In particular, Chapters 5 and 6 owe a great deal to Dr. A. B. Philpott, who proved many of the results given there and was responsible for a large part of the formulation of the algorithms. One of us (P. N.) would like to acknowledge the support provided by the Allen Clark Research Fellowship at Churchill College. Both of us are grateful for the facilities and the stimulating environment provided by the Management Studies Group of Cambridge University Engineering Department.

Eddie Anderson
Peter Nash

Contents

1

Infinite-dimensional Linear Programs

1.1 INTRODUCTION

A linear program is an optimization problem with a linear objective functional and linear constraints. In 1947, G. B. Dantzig discovered the simplex method for the solution of such problems, and since then the method has become perhaps the most widely used of all optimization techniques. Two main factors seem to have contributed to this popularity. On the one hand, the simplex method is observed to be very efficient in practice. On the other hand, the method provides a very complete solution to the problem. As well as the optimal values of the decision variables, the output of the simplex algorithm includes much information about the sensitivity of the optimal solution to changes in the problem data. This information can be very useful when the parameters of the problem are known only imprecisely. These factors have made it often worthwhile to accept the inaccuracies involved in using a linear model of a non-linear real-world problem.

Until comparatively recently, a complete theory of linear programming existed only for problems involving a finite number of decision variables subject to a finite number of constraints. We shall refer to such a problem as a finite or finite-dimensional LP. Formally, it is posed as

$$\begin{aligned} \text{FLP:}\quad &\text{minimize} && c^{\mathrm{T}} x \\ &\text{subject to} && Ax = b, \\ & && x \in R^n, \qquad x \geqslant 0, \end{aligned}$$

where $c \in R^n$, $b \in R^m$ and the $m \times n$ matrix A are given. The positivity constraint $x \geqslant 0$ is interpreted as meaning $x_i \geqslant 0$, $i = 1, 2, \ldots, n$.

It is difficult to date the first attempts to extend the theory of linear programming to more general settings, but an early example of a linear-programming problem posed other than in R^n was due to Bellman (1957), who examined a particular linear optimal control problem arising from a model of a production system. The control problem was posed as a linear program in a space of functions of a continuous-time variable. Such problems have come to be called continuous-time linear programs, and have been the object of considerable study, along with many other linear programs posed in a variety of abstract vector spaces. Our object here is to examine the theory of linear programming in this more general sense, and particularly as it relates to the extension to infinite-dimensional problems of the simplex and related algorithms.

Accordingly, this book is set out as follows. We begin by looking, in the remainder of this chapter, at some problems which can be posed as infinite-dimensional linear programs. We then examine briefly the elements of the simplex algorithm for FLP, and formulate some questions which must be addressed in any attempt to extend to the infinite-dimensional case the concepts which underlie the simplex algorithm. Examples demonstrate some of the difficulties involved in making these extensions. The resolution of some of these difficulties is the aim of Chapters 2 and 3. The remaining chapters of the book are then devoted to working out the theory and deriving algorithms for a number of particular classes of infinite-dimensional problems.

1.2 SOME INFINITE LINEAR PROGRAMS

To motivate our study, we examine here some problems which are naturally modelled as infinite linear programs.

1.2.1 The bottleneck problem (Bellman (1957))

In an economy, n different goods, $G_1, G_2, \ldots, G_n$, are produced by m different types of plant or production facility, $P_1, P_2, \ldots, P_m$. At the beginning of a five-year plan, there is available a certain capacity in each of these types of plant, and more can be made by re-investing the goods produced. The aim of the plan is to maximize the productive capacity at the end of the period.

Let $x_i(t), i = 1, 2, \ldots, m$, denote the rate of production of new capacity of type i at time t. Production of new plant requires the consumption of a certain quantity b_{ij} of good G_i for each additional unit of plant P_j. Thus the amounts of goods consumed in this way are given by $Bx(t)$, where $x(t)$ is the vector with components $x_1(t), x_2(t), \ldots, x_m(t)$ and B is the matrix whose i,jth element is b_{ij}. Let $z_i(t)$ denote the total productive capacity of type i available at time t. Denote by d_{ij} the rate of production of G_i for each unit of plant P_j. Then the total rates of production of goods at time t are given by $Dz(t)$, where D is the matrix whose i,jth element is d_{ij} and $z(t)$ is the vector $(z_1(t), z_2(t), \ldots, z_m(t))^{\mathrm{T}}$.

The constraint on investment in additional plant due to limitations in productive capacity is then given by

$$Bx(t) \leqslant Dz(t)$$

throughout the time period under consideration. If c_0 is the vector of initial productive capacities, we can write

$$z(t) = c_0 + \int_0^t x(\tau)\, \mathrm{d}\tau, \tag{1}$$

and hence

$$Bx(t) - \int_0^t Dx(\tau)\,\mathrm{d}\tau \leqslant c, \tag{2}$$

where $c = Dc_0$. If we wish to maximize a weighted sum, $\Sigma a_i z_i(T)$, of the production capacities at the end of the time period, then we obtain the following linear program

$$\begin{aligned} \text{BP:}\quad &\text{maximize} && \int_0^T a^{\mathrm{T}}\, x(t)\,\mathrm{d}t \\ &\text{subject to} && Bx(t) - \int_0^t Dx(\tau)\,\mathrm{d}\tau \leqslant c, \\ &&& x(t) \geqslant 0, \qquad t \in [0, T]. \end{aligned}$$

The decision variables here are the functions x_i. If we wish to maximize a weighted sum of the total production of the various goods over the whole time interval, then the form of the problem is very similar. An integration by parts is necessary and the vector a in the objective function is replaced by a function of time of the form $(1-t)\,D^{\mathrm{T}}a$.

This type of problem is called a continuous or continuous-time linear program. Various extensions to the basic form of the problem posed here have been studied, particularly the case where a, c, B and D vary with time. The name *bottleneck problem* refers to the constraint on production capacity.

1.2.2 Continuous-time network flow

A classical problem in finite linear programming is that of maximizing the flow of some commodity between two specified nodes (the *source* and the *sink*) in a transportation network whose arcs are subject to capacity limitations. When the capacity limitations do not vary with time the problem is solved by choosing a single set of constant flows, and this can be done by means of a simple algorithm due to Ford and Fulkerson (1962). This algorithm relies on an elegant duality theorem which is available for this particular problem. The duality theorem tells us that the problem of maximizing the flow in the network is equivalent to that of partitioning the network into two sets of nodes, one containing the source and the other containing the sink, in such a way as to minimize the total capacity of those arcs connecting nodes associated with the source to nodes associated with the sink. Such a partition is called a *cut*, and the duality theorem is sometimes called the *maximum-flow minimum-cut theorem.*

Infinite-dimensional versions of this problem arise in a number of contexts. Perhaps the most obvious is the case where the capacities of the arcs vary with time, and there is the possibility of storage at the nodes of the network. Consider, for example, a system of n reservoirs $R_1, R_2, \ldots, R_n$ from which a single, time-varying demand has to be met during some time period $[0, T]$. Suppose that the capacity of R_i is c_i, $i = 1, 2, \ldots, n$. Water flows into R_i at time t at a rate $r_i(t)$, and the demand at time t is $d(t)$. The maximum rate at which water can be fed from R_i to meet demand is f_i, a constant. If more water flows into any reservoir than it can hold, the rest is spilled to waste, and water may not be fed back into reservoirs. Subject to these constraints, we wish to meet as much as possible of the demand during the time interval $[0, T]$.

Let $x_i(t)$ denote the rate at which water is fed from R_i at time t. The feeder constraints can then be expressed

$$0 \leqslant x_i(t) \leqslant f_i, \qquad t \in [0, T], \quad i = 1, 2, \ldots, n. \tag{3}$$

Let $w_i(t)$ denote the rate of spillage from R_i at time t. The storage constraints are

$$0 \leqslant \int_0^t [r_i(\tau) - x_i(\tau) - w_i(\tau)] \, d\tau + s_i \leqslant c_i, \qquad t \in [0, T], \quad i = 1, 2, \ldots, n, \tag{4}$$

where s_i is the amount of water stored in R_i initially. The reservoir control problem can then be stated as

$$\text{maximize} \quad \int_0^T \left[\sum_{i=1}^n x_i(t) \right] dt$$

subject to (3) and (4), and the further constraints that the total rate of feed is no greater than the demand, that is

$$\sum_{i=1}^n x_i(t) \leqslant d(t). \qquad t \in [0, T],$$

and that

$$w_i(t) \geqslant 0, \qquad t \in [0, T], \quad i = 1, 2, \ldots, n.$$

This problem is obviously a special case of the bottleneck problem of the previous section. To see how it can be posed as a continuous-time network-flow problem, consider the capacitated network $\Omega(t)$ shown in Figure 1.1. This has a storage node representing each reservoir. An arc of capacity f_i connects R_i to a single node D_2 of zero storage capacity, which is in turn connected to the sink B_t by an arc of capacity $d(t)$. Conservation of flow at this zero-storage node ensures that the total feed is no greater than $d(t)$. Inflows are represented by an infinite-capacity arc connecting the source A_t to the infinite-storage node D_1, which is connected by arcs of capacity $r_1(t), r_2(t), \ldots, r_n(t)$ to the reservoirs. Spillage is represented by flows in infinite capacity arcs connecting the reservoirs to D_1.

Continuous-time network flow problems in which there is no storage are an essentially straightforward extension of the static case. The solution to such a problem is obtained by maximizing the instantaneous flow at each time. Assuming that the problem data are well enough behaved to allow us to choose these instantaneous flows appropriately, then the maximum flow over a time interval is just the integral of the instantaneous maxima. Of course, the problem cannot be solved like this in practice, as the number of flow maximizations involved is infinite. In reality, one uses parametric programming techniques to

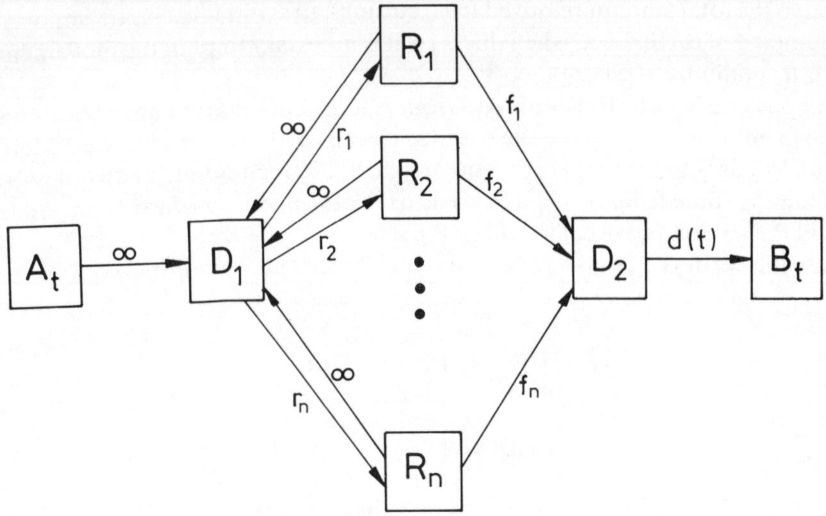

Figure 1.1 The network $\Omega(t)$ for the reservoir control problem

find the times when the maximum-flow regime changes, and relies on the good behaviour of the problem data to ensure that these are only finite in number.

The presence of non-zero storage capacities complicates the problem a good deal. The effect of storage is to link flows at different times. In a discrete-time formulation, this can be dealt with by considering a number of copies (one per time epoch) of the basic network, with directed arcs to represent storage capacity linking successive copies of each node at which storage is allowed. The result of this construction is an augmented network for which the maximum flow can be computed by the Ford–Fulkerson algorithm, and this provides the solution to the original problem.

One approach to the continuous-time problem is to approximate it using a discrete-time formulation. While this is clearly a workable method, the augmented network will be very large if a fine discretization is needed. Moreover, one feels that the discretization may hide essential structural features of the solution, and it is natural to ask whether a treatment in continuous time is possible. We shall see in Chapter 6 that the Ford–Fulkerson algorithm can be extended to the continuous-time problem.

1.2.3 Cutting and filling

A road built on undulating terrain does not usually follow exactly the contours of the land, but is supported on embankments and run through cuttings, so as to reduce the number and severity of the gradients. In building the road, large masses of earth have to be moved to create the cuttings and embankments. It is advantageous if the total amount of earth needed for embankments is roughly equal to the total amount removed from cuttings, to avoid large-scale importation or dumping of earth. Even when this is so, it is necessary to plan the movements of earth to minimize their total cost.

Suppose we have a stretch of road from A to B. Let s denote distance along the route from A, and let $\Psi(s)$ be the planned elevation of the road above the terrain. Let $c(s, s')$ denote the unit cost of moving earth between points at distances s and s' along the route from A. Let $\pi(s, s')\,\mathrm{d}s\,\mathrm{d}s'$ be the quantity of earth moved from within the section $(s, s+\mathrm{d}s)$ to within the section $(s', s'+\mathrm{d}s')$. Then the total cost of moving earth is $\int\int c(s, s')\pi(s, s')\,\mathrm{d}s\,\mathrm{d}s'$. This has to be minimized, subject to

$$\int \pi(s, s')\,\mathrm{d}s = \Psi_1(s') \qquad \text{for all } s',$$

$$\int \pi(s, s')\,\mathrm{d}s' = \Psi_2(s) \qquad \text{for all } s,$$

$$\pi(s, s') \geqslant 0 \qquad \text{for all } s, s',$$

where

$$\Psi_1(s) = \max\{\Psi(s), 0\},$$

$$\Psi_2(s) = \max\{-\Psi(s), 0\}.$$

Similar 'cutting and filling' problems arise in levelling a plot of land, or moving a mass of earth in three dimensions to form an earthwork. The study of such problems has a long history, actually initiated in the study of military operations involving earthworks. As is apparent, the problem is an infinite-dimensional version of the well-known transportation problem of linear programming.

We shall give a full description of this problem in Chapter 5. Here we just note that the problem in more general form is

$$\begin{aligned} &\text{minimize} && \int_{X\times Y} c(x, y)\,\mathrm{d}\pi \\ &\text{subject to} && P_1\pi = \rho_1, \\ & && P_2\pi = \rho_2, \\ & && \pi \geqslant 0, \end{aligned}$$

where ρ_1 and ρ_2 are prescribed measures on two sets X and Y; π is a measure on $X\times Y$, to be found; P_1 and P_2 are the operators which project measures on $X\times Y$ onto measures on X and Y respectively; and $c(x, y)$ is a given continuous function on $X\times Y$. The form of the problem stated originally occurs when π is an absolutely continuous measure, so that for any measurable set $S \subset X\times Y$

$$\pi(S) = \int_S \pi(x, y)\,\mathrm{d}x\,\mathrm{d}y.$$

1.2.4 Approximation

The three examples we have just discussed all involve linear programs in which both the decision variables and the constraints lie in infinite-dimensional spaces. An interesting extension of linear programming, in which the number of variables remains finite but there are an infinite number of constraints, arises from a large class of problems of approximation. Such problems are called semi-infinite linear programs.

Suppose we wish, perhaps for purposes of computation, to approximate some function by a linear combination of the members of a finite set of simpler functions, for example by a polynomial of given degree. That is, given a set $\{\psi_1, \psi_2, \ldots, \psi_n\}$ of functions, we wish to find a function ψ in the form

$$\psi = \sum_{i=1}^{n} \alpha_i \psi_i$$

which best approximates some given function f over some interval $[x_1, x_2]$. The goodness of fit of our approximation could be measured in various ways; one way is by the maximum absolute error. If we do this, the approximation problem may be formulated

$$\begin{aligned} &\text{minimize} \quad && r \\ &\text{subject to} \quad && -r \leqslant f(x) - \sum_{i=1}^{n} \alpha_i \psi_i(x) \leqslant r, \qquad x \in [x_1, x_2]. \end{aligned}$$

This problem has an infinite number of constraints on the $n+1$ variables r, α_1, $\alpha_2, \ldots, \alpha_n$. We return to a discussion of problems like this in Chapter 4.

1.2.5 Games

There is an intimate connection between linear programming and certain types of two-player game, well-known in the finite-dimensional case. If we look at games of this type, but with the feature that the players choose strategies from infinite-dimensional spaces, we would expect to find the same connection with infinite-dimensional linear programs. The following is an example.

Suppose that two people take part in a duel. Each may fire only one shot, and their weapons are so efficiently silenced that neither can tell whether or not his opponent has fired. A shot fired by either has a probability of hitting its target which varies with time, perhaps as the duellists' nerves are affected by tension. The duellists are equally skilled as marksmen. Both are mathematicians (the duel is about priority in the discovery of an elegant theorem in number theory), and are concerned as to how to choose their moment to fire.

The determination of an optimal strategy for the duellists is a well-known problem in game theory. If the duel starts at time 0 and, and is declared a draw if neither duellist is hit by time 1, then we have an example of what is usually called a two-person game on the unit square. To formulate the problem as a linear program, let $h(t)$ denote the probability that a shot fired at time t hits its target.

The usual minimax strategy involves each duellist maximizing his chances of winning, given the most pessimistic assumptions about his opponent's strategy. Let us look at this from the standpoint of one of the duellists, called player 1. If he chooses a strategy such that the time of firing of his shot is distributed with probability measure μ on [0, 1], and his opponent, player 2, chooses a corresponding probability distribution ν, then the probability that player 1 wins the duel is

$$p_1(\mu, \nu) = \int_0^1 \left\{ \int_t^1 h(t)\,\mathrm{d}\nu(s) \right\} \mathrm{d}\mu(t) + \int_0^1 \left\{ \int_0^t (1-h(s))\,h(t)\,\mathrm{d}\nu(s) \right\} \mathrm{d}\mu(t),$$

while the probability that player 2 wins is

$$p_2(\mu, \nu) = \int_0^1 \left\{ \int_0^t h(s)\,\mathrm{d}\nu(s) \right\} \mathrm{d}\mu(t) + \int_0^1 \left\{ \int_t^1 (1-h(t))\,h(s)\,\mathrm{d}\nu(s) \right\} \mathrm{d}\mu(t),$$

and with probability $1-p_1(\mu, \nu)-p_2(\mu, \nu)$ neither player is hit. If we take the value obtained by a player as 1 if he wins, -1 if he loses and 0 otherwise, then the expected value to player 1 is $p_1(\mu, \nu)-p_2(\mu, \nu)$. To determine the minimax strategy, we need to solve

$$\max_{\mu \in P} \; \min_{\nu \in P} \{p_1(\mu, \nu) - p_2(\mu, \nu)\}$$

where P is the set of probability measures on [0, 1].

Now for any fixed μ, the inner minimization takes the form

$$\text{minimize} \int_0^1 f_\mu(s)\,\mathrm{d}\nu(s),$$

and takes the value $\min\{f_\mu(s)\colon s \in [0, 1]\}$, which is attained when ν has unit mass at a point where f_μ is minimized. Here,

$$f_\mu(t) = \int_0^t \{h(\tau)-h(t)+h(\tau)h(t)\}\,\mathrm{d}\mu(\tau) + \int_t^1 \{h(\tau)-h(t)-h(\tau)h(t)\}\,\mathrm{d}\mu(\tau).$$

From this we obtain the linear program

$$\begin{aligned} &\text{maximize} && z \\ &\text{subject to} && z \leqslant f_\mu(s), \quad s \in [0, 1], \\ & && \mu \in P. \end{aligned}$$

In other words,

$$\begin{aligned} &\text{maximize} \quad z \\ &\text{subject to} \quad z - \int_0^1 g(s, t)\,\mathrm{d}\mu(t) \leqslant 0, \qquad s \in [0, 1], \\ &\int_0^1 \mathrm{d}\mu(t) = 1, \\ &\mu \geqslant 0, \end{aligned}$$

where

$$g(s, t) = h(t) - h(s) + h(s)h(t) \qquad \text{if } t \leqslant s,$$

$$g(s, t) = h(t) - h(s) - h(s)h(t) \qquad \text{if } t > s.$$

1.3 ELEMENTS OF THE SIMPLEX METHOD

If we wish to solve the various infinite-dimensional problems that we have been describing, then we are naturally led to ask whether there is any way of extending the simplex algorithm, and other related algorithms, to such problems. The simplex method for linear programming is based on three main ideas: duality, basic solutions and pivoting. Here we examine briefly the roles of these ideas in finite linear programming, and give some examples to show that their extension to infinite-dimensional problems is not straightforward.

1.3.1 Duality

With the finite-dimensional linear program, FLP, can be associated a dual problem:

$$\begin{aligned} \text{FLP*:} \quad & \text{maximize} && b^{\mathrm{T}} y \\ & \text{subject to} && A^{\mathrm{T}} y \leqslant c, \\ &&& y \in R^m . \end{aligned}$$

The properties of FLP and FLP* are intimately linked. Their relationship is summarized in the duality theorem of finite linear programming, which we now state.

Theorem 1.1

(a) *If x and y are feasible solutions for* FLP *and* FLP* *respectively, then* $c^{\mathrm{T}} x \geqslant b^{\mathrm{T}} y$.

(b) *If x^* is optimal for* FLP, *then there exists y^* which is optimal for* FLP* *and* $c^{\mathrm{T}} x = b^{\mathrm{T}} y$.

(c) *If x^* and y^* are optimal for* FLP *and* FLP* *respectively, then* $x^{*\mathrm{T}} (c - A^{\mathrm{T}} y^*) = 0$.

(d) *If both* FLP *and* FLP* *have feasible solutions, then they both have optimal solutions.*

Part (a) of this theorem is often called *weak duality*; part (b) is strong duality. The condition in part (c) is called *complementarity* or *complementary slackness*. If x^* is feasible for FLP and y^* is feasible for FLP*, then both x^* and $(c - A^{\mathrm{T}} y^*)$ are non-negative vectors, and so condition (c) is equivalent to the statement that for each i, either $x_i^* = 0$ or $(c - A^{\mathrm{T}} y^*)_i = 0$.

It is the property of strong duality which provides the possibility of verifying in a simple way the optimality of a given solution x^* for FLP. Suppose a solution y^* can be found for FLP*, such that $c^{\mathrm{T}} x^* = b^{\mathrm{T}} y^*$. Then, by virtue of (a) of the theorem, x^* and y^* are both optimal. On the other hand, if x^* is optimal, such a y^* can always be found, because of property (b). This would not necessarily be very helpful, in the absence of any method of constructing y^*. However, a second crucial property of finite linear programs is that an optimal solution, where one exists, can always be found among a restricted set of *basic* solutions. If x^* is a basic solution, there is a simple construction which generates y^*. The role of basic solutions is examined below, but first we will look at an example of a continuous-time linear program and its dual. This will give some indication of the difficulties that will confront us when we come to extend the notion of duality to the infinite-dimensional case.

Consider the problem

$$\text{P1:}\quad \text{maximize} \quad \int_0^1 x_2(t)\,dt$$

$$\text{subject to} \quad x_1(t) \leqslant 2, \tag{5}$$

$$x_2(t) \leqslant 4, \tag{6}$$

$$0 \leqslant \int_0^t \{x_1(s) - x_2(s)\}\,ds \leqslant 1, \tag{7}$$

$$x_1(t),\ x_2(t) \geqslant 0, \qquad t \in [0, 1].$$

One way to view this problem is as a very simple continuous-time network flow problem, for which the network $\Omega(t)$ is shown in Figure 1.2, with maximum storage of one unit at R. The problem P1 has many optimal solutions, one of which is

$$x_1(t) = 2, \quad t \in [0, 1],$$

$$x_2(t) = 0, \quad t \in [0, \tfrac{1}{2}],$$

$$x_2(t) = 4, \quad t \in [\tfrac{1}{2}, 1].$$

Thus the objective function of P1 has an optimal value of 2.

There are three ways in which we could go about constructing the dual of a given linear program: we can proceed purely algebraically, looking at an adjoint system of constraints, or we can formulate a Lagrangian for the problem and deduce the form of the dual from this, or we can look at a sequence of finite-dimensional approximations, perhaps via a sequence of discretizations, and look at the sequence of dual problems in the limit. Adopting the third approach,

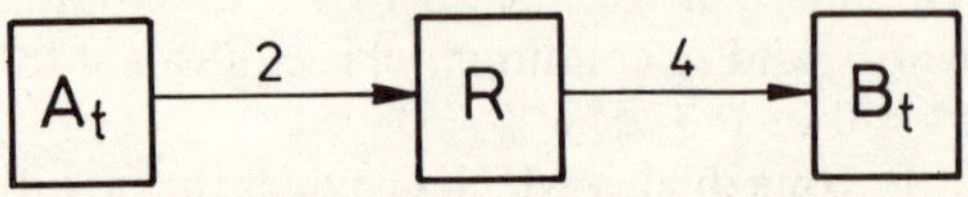

Figure 1.2 The network $\Omega(t)$ for problem P1

consider a sequence of finite-dimensional approximations P1 (n) to P1, obtained by discretization.

$$\text{P1}(n)\text{:}\quad \text{maximize}\quad (1/n)\sum_{i=1}^{n} x_2(i)$$

$$\text{subject to}\quad x_1(i) \leqslant 2,$$

$$x_2(i) \leqslant 4,$$

$$0 \leqslant (1/n)\sum_{j=1}^{i} [x_1(j) - x_2(j)] \leqslant 1,$$

$$x(i) \geqslant 0, \qquad i = 1, 2, \ldots, n.$$

The dual of this finite-dimensional linear program is

$$\text{P1}(n)^*\text{:}\quad \text{minimize}\quad \sum_{i=1}^{n} [2y_1(i) + 4y_2(i) + w_2(i)]$$

$$\text{subject to}\quad y_1(i) - (1/n)\sum_{j=i}^{n} [w_1(j) - w_2(j)] \geqslant 0,$$

$$y_2(i) + (1/n)\sum_{j=i}^{n} [w_1(j) - w_2(j)] \geqslant 1/n,$$

$$y(i) \geqslant 0, \quad w(i) \geqslant 0, \qquad i = 1, 2, \ldots, n.$$

After multiplying the constraints of P1(n)* by n, and setting $\mu(i) = ny(i)$, $\lambda(i) = nw(i)$, $i = 1, 2, \ldots, n$, P1(n)* becomes

$$\text{minimize}\quad (1/n)\sum_{i=1} [2\mu_1(i) + 4\mu_2(i) + \lambda_2(i)]$$

$$\text{subject to}\quad \mu_1(i) - (1/n)\sum_{j=i}^{n} [\lambda_1(j) - \lambda_2(j)] \geqslant 0,$$

$$\mu_2(i) + (1/n)\sum_{j=i}^{n} [\lambda_1(j) - \lambda_2(j)] \geqslant 1,$$

$$\mu(i) \geqslant 0, \quad \lambda(i) \geqslant 0, \qquad i = 1, 2, \ldots, n.$$

Taking a formal limit of the sequence $\{\text{P1}(n)^*\}$, we arrive at the problem P1*

$$\text{P1}^*\text{:}\quad \text{minimize}\quad \int_0^1 [2\mu_1(t) + 4\mu_2(t) + \lambda_2(t)]\, dt$$

$$\text{subject to}\quad \mu_1(t) - \int_t^1 [\lambda_1(s) - \lambda_2(s)]\, ds \geqslant 0, \tag{8}$$

$$\mu_2(t) + \int_t^1 [\lambda_1(s) - \lambda_2(s)]\, ds \geqslant 1, \tag{9}$$

$$\mu(t) \geqslant 0, \quad \lambda(t) \geqslant 0, \qquad t \in [0, 1].$$

P1* is dual to the original problem P1; the reader may like to verify that the same dual formulation is obtained by either of the other two approaches mentioned above. For the moment, we simply explore the formal properties of P1*, leaving aside obvious questions about the validity of the sort of limiting process we used to derive it.

Suppose we seek an optimal solution μ^*, λ^* to this dual problem, complementary with our chosen optimal solution to P1. The natural interpretation here of complementarity is that the optimal dual functions λ^* and μ^* are non-zero only when the appropriate primal constraints bind. The integral constraints (7) in P1 are binding only at $t = 0$, $t = \frac{1}{2}$, and $t = 1$, so the complementarity condition suggests that

$$\lambda_1^*(t) = \lambda_2^*(t) = 0,$$

except when $t \in \{0, \frac{1}{2}, 1\}$. Moreover, the second capacity constraint (6) is not binding in $[0, \frac{1}{2}]$, which suggests that $\mu_2(t) = 0$ for $t \in [0, \frac{1}{2}]$. This contradicts the second constraint (9) of P1*.

In fact, in the setting implied by our notation, in which λ and μ are functions, P1 and P1* are not strongly dual. It is not possible to find measurable functions λ and μ which achieve the objective function value 2 in P1*. This value can be approached arbitrarily closely, and if we allow λ to be chosen from a larger space than that of measurable functions, in particular if we allow λ to include δ-functions, then it is achievable. Thus the dual problem has the optimal 'solution'

$$\lambda_1(t) = 0, \quad t \in [0, 1],$$

$$\lambda_2(t) = \delta(t-1),$$

$$\mu_1(t) = 1, \ t \in [0, 1],$$

$$\mu_2(t) = 0, \ t \in [0, 1].$$

It is easily seen that this does indeed achieve an objective function value of 2. Since P1 and P1* are weakly dual, as is easily verified, this is an optimal solution for P1* in this wider space.

A crucial point made by this example is that there may not be a single obvious dual program for a given infinite-dimensional primal, since there may well be a choice of spaces in which to pose the dual (and for that matter, the primal itself), and the particular choice that is made affects the type of duality relation which holds. The theory of duality for infinite-dimensional linear programs is therefore concerned with establishing conditions under which strong duality holds for at least some choice of spaces, and with the investigation of when strong duality holds with primal and dual posed in 'nice' spaces. When these conditions cannot be guaranteed, one becomes interested in extended notions of duality which may hold even when conventional strong duality does not.

1.3.2 Basic solutions and pivoting

Fundamental to the simplex algorithm for linear programming is the fact that whenever FLP has an optimal solution, then one can be found among the extreme points of the set F of feasible solutions (an extreme point of a set is a member of the set which is not an interior point of any line segment lying in the set). The extreme points of F can be characterized as the basic feasible solutions. An element $x \in F$ is a basic feasible solution if the number of non-zero components of x is no greater than the rank of A; x is called non-degenerate if this number is equal to the rank of A, and degenerate otherwise.

Denote by $S(x)$ the set whose elements are the indices of the non-zero components of some basic feasible solution x; $S(x)$ is called the *basis*. As x is basic, the definition given above implies that x is the unique solution of $A\xi = b$ for which ξ_i is non-zero only if $i \in S(x)$. Moreover, when x is non-degenerate, there is a unique pair (y, w) which satisfies

$$A^{\mathrm{T}}y + w = c, \tag{10}$$

$$(c - A^{\mathrm{T}}y)_i = 0, \qquad i \in S(x). \tag{11}$$

This happens because the matrix formed by the set of columns of A indexed by $S(x)$ is invertible, so that (11) determines y, and w is then determined by (10).

Now if x is a basic optimal solution, the solution to (11) must be the optimal dual solution, since (11) is just the complementarity condition of Theorem 1.1. Feasibility of the optimal dual solution then implies that

$$A^{\mathrm{T}}y \leqslant c,$$

or in other words that $w \geqslant 0$. On the other hand, if (y, w) is a solution to (10) and (11) constructed from a basic feasible solution x, and $w \geqslant 0$, then y is feasible for FLP*. Furthermore, (11) implies that

$$b^{\mathrm{T}}y = x^{\mathrm{T}}A^{\mathrm{T}}y = c^{\mathrm{T}}x,$$

and so y is optimal for FLP* and x is optimal for FLP.

These properties allow us to check the optimality of any given basic feasible solution x for FLP, by constructing (y, w) to satisfy (10) and (11). If $w \geqslant 0$, then x is optimal. Otherwise, an improved solution can be generated by an appropriate change to x. The constructed value of w is a reduced gradient, indicating feasible directions of perturbation which will produce an improvement in objective function value. That is, for any other feasible solution ξ,

$$c^{\mathrm{T}}x - c^{\mathrm{T}}\xi = w^{\mathrm{T}}(x - \xi).$$

Since w and x are complementary, we need look only at those components of ξ whose indices are outside the basis in evaluating this last expression. Hence the negative components of w correspond to currently zero components of x which should be increased from zero in our next, improved solution. Feasibility is maintained by making appropriate changes to the components of x which are in the current basis. By choosing to increase just one of the non-basic components, and continuing to increase it until some other component, currently in the basis, reaches zero, we will arrive at a new basic feasible solution (or, if no currently basic component ever becomes zero, we will decrease the objective function value without limit).

The mechanism for selecting exactly which component is to be increased, and generating the new solution, is called pivoting. More than one pivoting method has been used in simplex and related algorithms, but in each case the underlying mechanism depends on the specific algebraic properties of the constraint equations and the underlying spaces. As we shall see, deriving analogues of these mechanisms is perhaps the most difficult task of all in developing a simplex-like algorithm for an infinite-dimensional problem. We will not dwell on this subject here, but return to a discussion of the basic solutions themselves.

The fundamental property of a basic solution x is that the equation $Ax = b$ has exactly one solution in the space $B(x)$, defined by

$$B(x) = \{\xi : \xi \in R^n, \xi_i = 0 \text{ for } i \in S(x)\}.$$

In order to see what might happen when we try to generalize this notion to infinite-dimensional problems, we consider an example, due to Perold (1981), of a simple continuous-time constraint:

$$3tx(t) - 6\int_0^t x(s)\,ds = t^3,$$

$$x(t) \geqslant 0, \quad t \in [0, 1].$$

In, say, the space of continuous functions on [0, 1], this constraint defines the feasible region $\{\xi_\alpha : \alpha \in R, \alpha \geqslant 0\}$, where

$$\xi_\alpha(t) = t^2 + \alpha t.$$

It is immediate that this set has a single extreme point, namely ξ_0.

Proceeding by analogy with the finite case, one might reason that the place of the basis should in this case be taken over by the support of ξ, and be led to define

$$S(x) = \{t : t \in [0, 1], x(t) > 0\},$$
$$B(x) = \{\xi : \xi(t) = 0 \text{ for } t \in S(x)\}.$$

If we apply these definitions to the above example, with $x = \xi_0$, we find that $\xi_\alpha \in B(x)$ for all $\alpha \geqslant 0$. Thus the most natural-seeming extension of the definition of basic solutions, as being uniquely determined by their support, will not correspond to the extreme points.

In the next chapter, we will begin our study of infinite-dimensional linear programs by setting up for a completely abstract program a definition of the basic solutions. This definition is an algebraic restatement of the definition of the extreme points of the feasible set. We also define degeneracy of basic solutions in an abstract setting. Non-degenerate basic feasible solutions, where they exist, have all the properties they exhibit in the finite-dimensional case, but we shall see that degeneracy is a much more important phenomenon in infinite-dimensional linear programming than in the finite-dimensional case.

Problems

1.1 Find an optimal solution for the problem P1(n)* of Section 1.3.1, and show how this approaches a solution involving a δ-function as n approaches infinity.

1.2 Show that in the silent duel as formulated in Section 1.2.5, the problem faced by player 2 can be identified with an LP dual of player 1's problem.

1.3 Many constraint sets of the type given in Perold's example do not exhibit the slightly unexpected behaviour of that example. Consider the constraints

$$\int_0^t x_1(s)\,ds + x_2(t) = t,$$

$$x_2(t) + x_3(t) = 1,$$

$$x_1(t) \geqslant 0, \quad x_2(t) \geqslant 0, \quad x_3(t) \geqslant 0, \qquad t \in [0, 1].$$

Find a solution which is an extreme point of the feasible set, and show that it is determined by its support.

2

Algebraic Fundamentals

2.1 INTRODUCTION

We have seen that the underlying concepts of linear programming in finite-dimensional spaces may not necessarily extend in a straightforward way to the infinite-dimensional case. Some at least of the features of the theory for the finite-dimensional problem must depend on the properties of R^n. At the same time, only a very small amount of structure is required to allow a linear program to be posed. In fact, apart from the linear vector space structure, the formulation of a linear program requires only that the notion of positivity—that is, an order structure of some sort—should be defined.

We begin our study by postulating this bare minimum of structure. Our purpose is to discover what properties of linear programs stem from this, independently of other structural properties of the spaces in which the problem variables lie. We make use only of the algebraic properties of the problem, and properties arising out of the order structure. In the next chapter, we will go much more deeply into some aspects of the theory, particularly duality, by postulating a topological structure and employing the tools of functional analysis, but here our results are algebraic. In many cases, they are derived simply by noticing that once the correct definitions are used, many of the manipulations available in the finite-dimensional case carry over to an abstract setting. Thus the ideas presented in this chapter are not particularly hard to grasp, and the theorems are relatively easy to prove. However, the two crucial definitions, of basic solutions and reduced costs, will turn out to be essential starting points for the specification of simplex-like algorithms for the infinite-dimensional linear programs which we consider in later chapters.

2.2 LINEAR PROGRAMS AND THEIR ALGEBRAIC DUALS

Let X and Z be real linear vector spaces. Let P be a convex cone in X (a convex cone is a set closed under addition and multiplication by positive scalars). Let X be partially ordered by the relation $\leqslant$, defined by

$$x \leqslant y \quad \text{if } y - x \in P, \quad (x, y \in X).$$

We write $y \geqslant x$ if $x \leqslant y$. In any vector space, we denote the null vector by θ.

Clearly, $x \in P$ if and only if $x \geqslant \theta$, so we call such an x *positive*. P is called the *positive cone* in X. Note that at this stage we assume nothing about the positive cone other than its convexity. In particular, we do not assume that P contains no subspace of X. Thus in general, $x \leqslant y$, $y \leqslant x$ do not together imply $x = y$. The assumption that P contains no subspace will at times be made, but in developing the theory we shall quite often wish to take P to be the whole of X.

A *linear functional* on X is a linear map from X to the real line. The set of all such maps, with the operations of addition and scalar multiplication defined in the obvious way, is a vector space, called the *algebraic dual* of X and denoted by X^*. If $x \in X$ and $c^* \in X^*$, then we denote the image of x under c^* by $\langle x, c^* \rangle$. A positive cone P^* in X^* may be defined by

$$P^* = \{x^* \in X^* : \langle x, x^* \rangle \geqslant 0 \text{ for all } x \in P\},$$

and there is an associated partial ordering on X^*, again denoted by $\leqslant$.

2.2.1 The linear program

Let A be a linear map from X to Z, b an element of Z and let c^* be a linear functional on X. The primal linear program is

$$\begin{aligned} \text{LP:} \quad \text{minimize} \quad & \langle x, c^* \rangle \\ \text{subject to} \quad & Ax = b, \\ & x \in X,\ x \geqslant 0. \end{aligned}$$

This is an equality-constrained program; to pose an inequality-constrained program, one must assume the existence of a positive cone and associated partial order in Z, and replace the constraint $Ax = b$ with the constraint $Ax \geqslant b$. In this chapter, we shall deal mainly with the equality-constrained program. All the results can be carried over to the inequality-constrained problem by the usual procedure of adding slack 'variables' to transform the inequality constraints to equality form. Thus the constraint $Ax \geqslant b$ becomes $Ax - z = b$, $z \geqslant \theta$, $z \in Z$. Sometimes, a little care is needed in carrying over results from one case to the other, but we leave the details to the reader.

For any program P, we say that x is *feasible* if x satisfies the constraints of P, including any positivity constraints. The set of feasible solutions for P is called F (P), or just F when the context makes it clear which program is referred to. Thus we have

$$F(\text{LP}) = \{x : x \geqslant \theta,\ Ax = b\}.$$

A program for which at least one feasible solution exists is called *consistent*.

Example 1. *The finite linear program.* The finite linear program was introduced as FLP in Chapter 1. For FLP, X is R^n, and P is the set of vectors all of whose components are non-negative. X^* is also R^n, and $\langle x, c^* \rangle$ is just the inner product $c^{\mathrm{T}}x$. The space Z is some R^m, and the map A is represented by an $m \times n$ matrix.

Example 2. The bottleneck problem. In the bottleneck problem introduced in the previous chapter, we could take X to be the space of continuous functions on $[0, T]$ taking values in R^m, and positive cone $P \subset X$ to be defined by

$$P = \{x : x(t) \geqslant 0,\ t \in [0, T]\}.$$

The objective functional c^* is defined by

$$\langle x, c^* \rangle = \int_0^T a^{\mathrm{T}} x(t)\, \mathrm{d}t.$$

We can take Z to be the space of functions on $[0, T]$ taking values in R^n, and the map A is defined by

$$(Ax)\,(t) = Bx(t) - \int_0^t Dx(s)\, \mathrm{d}s.$$

This is an example of an inequality-constrained infinite-dimensional problem.

2.2.2 The algebraic dual problem

Let X^* and Z^* be the algebraic dual spaces of X and Z respectively. Let A^* be the map from Z^* to X^* defined by

$$\langle x, A^* z^* \rangle = \langle Ax, z^* \rangle, \qquad \text{for all } x \in X,\ z^* \in Z^*.$$

A^* is the *adjoint* (sometimes called the *transpose*) of A. We associate with LP an *algebraic dual* problem,

$$\begin{aligned} \text{AD:} \quad & \text{maximize} && \langle b, z^* \rangle \\ & \text{subject to} && A^* z^* \in c^*, \\ & && z^* \in Z^*. \end{aligned}$$

A weak duality relation between LP and AD is readily established.

Theorem 2.1 (*Weak duality*) *If x is feasible for* LP *and z^* is feasible for* AD, *then*

$$\langle x, c^* \rangle \geqslant \langle b, z^* \rangle.$$

Equality holds only if x and z^* are optimal solutions to LP *and* AD *respectively, and then*

$$\langle x, A^* z^* - c^* \rangle = 0.$$

Proof. Since x and z^* are feasible for LP and AD respectively,

$$\begin{aligned} \langle x, c^* \rangle &\geqslant \langle x, A^* z^* \rangle \\ &= \langle Ax, z^* \rangle \\ &= \langle b, z^* \rangle. \end{aligned}$$

Clearly, equality occurs only if $\langle x, A^* z^* - c^* \rangle = 0$. □

The solutions of LP and AD are related in another way. Suppose x and x' are feasible for LP, and that $(\xi^*, \zeta^*) \in X^* \times Z^*$ is any solution of the equation

$$A^* \xi^* + \zeta^* = c^*.$$

Then ζ^* is a reduced gradient for the primal problem, in the sense that

$$\langle (x - x'), c^* \rangle = \langle (x - x'), \zeta^* \rangle.$$

This follows from the fact that $x - x'$ lies in the null space of A, while $\zeta^* - c^*$ lies in the range of A^*. Thus

$$\begin{aligned}\langle (x - x'), (\zeta^* - c^*) \rangle &= \langle (x - x'), A^* \xi^* \rangle \\ &= \langle A(x - x'), \xi^* \rangle \\ &= 0.\end{aligned}$$

The simplex algorithm for the finite-dimensional linear program exploits these relationships by constructing at each step a basic solution for the primal problem, and a complementary solution for the dual constraint equations. If this dual solution is feasible (that is, if it satisfies the positivity constraints as well as the constraint equations), then the primal solution is optimal (as is the dual), by the weak duality theorem. If the dual solution does not satisfy the positivity constraints, then since it is a reduced gradient, it indicates directions of perturbation that will improve the objective function value. Crucial to these constructions are the properties of basic solutions, and we now examine these in an abstract setting.

2.3 BASIC SOLUTIONS

Let us review the important properties of basic solutions for the finite-dimensional linear program. These are: first that the basic solutions are just the extreme points of the set of feasible solutions; second, that, by restricting attention to vectors which are zero outside the basis, the constraint equations become uniquely solvable; and third, that a non-degenerate basic feasible solution defines a unique complementary dual solution. In looking for an abstraction of the definition of basic feasible solutions, we seek to reproduce these properties as far as possible. For the present, we leave aside the more fundamental question of whether basic solutions actually exist for particular linear programs.

For any $x \in P$, define

$$B(x) = \{\xi \in X : x + \lambda\xi \in P,\ x - \lambda\xi \in P \text{ for some scalar } \lambda > 0\}.$$

A feasible solution x for LP is *basic* if $B(x) \cap N(A) = \{\theta\}$, where we write $N(A)$ for the null space of A. The set $B(x)$ is clearly a linear subspace of X, and $x \in B(x)$. It is readily seen that when X is R^n, then this definition of $B(x)$ reduces to that given in Section 1.3.2, namely

$$B(x) = \{\xi \in R^n\colon \xi_i = 0 \text{ for } i \text{ not in } S(x)\},$$

where $S(x)$ is the support of x (often called the basis).

This definition of a basic solution is just an algebraic restatement of the definition of an extreme point of the feasible set, as the following theorem shows.

Theorem 2.2 *x is a basic feasible solution for* LP *if and only if x is an extreme point of F.*

Proof. A feasible solution x is not an extreme point of F if and only if there exists $\lambda > 0$ and $\xi \neq \theta$ such that $x + \lambda\xi$ and $x - \lambda\xi$ are both feasible. This is true if and only if

$$x + \lambda\xi \in P,$$

$$x - \lambda\xi \in P,$$

and

$$A(x + \lambda\xi) = A(x - \lambda\xi) = Ax = b;$$

that is, if and only if $\xi \in B(x) \cap N(A)$. □

It is straightforward to show that this definition of basic solutions reduces to the standard definition in the case of finite-dimensional LP: there, $B(x)$ is just the subspace of vectors whose components, other than those indexed by the basis, are zero, and the condition $B(x) \cap N(A) = \{\theta\}$ simply says that the columns of A indexed by the basis are linearly independent.

As an illustration of the way the definition works in the infinite-dimensional case, we look again at the example of Perold discussed in Chapter 1. Consider the set of continuous functions on [0, 1] which satisfy

$$(A\xi)(t) = 3t\xi(t) - 6\int_0^t \xi(s)\,\mathrm{ds} = t^3, \qquad t \in [0, 1], \qquad \xi \geqslant \theta,$$

where the positive cone in the space of continuous functions on [0, 1] is defined in the usual way, by $P = \{\xi : \xi(t) \geqslant 0,\ t \in [0, 1]\}$. For these constraints,

$$F = \{\xi_\alpha : \xi_\alpha(t) = t^2 + \alpha t,\ \alpha \in R,\quad \alpha \geqslant 0,\ t \in [0, 1]\}.$$

F has just a single extreme point, namely ξ_0: while

$$N(A) = \{\xi : \xi(t) = \alpha t,\ \alpha \in R\}.$$

Suppose we define the support of x, $S(x) = \{t : x(t) > 0\}$, and, as seems natural, try to use this as an analogue of the basis in finite LP. This would lead us to expect A to be invertible on the subspace $B'(x)$, defined by

$$B'(x) = \{\xi : \xi(t) = 0 \text{ for all } t \text{ not in } S(x)\}.$$

However, for this particular example, $B'(\xi_0)$ actually contains $N(A)$, so as we saw in Chapter 1, this approach will not work. On the other hand, $B(\xi_0)$ is the subspace of $C[0, 1]$ consisting of functions f with the property that $|f(t)| \leqslant Mt^2$, $t \in [0, 1]$, for some M. Clearly, $B(\xi_0) \cap N(A) = \{\theta\}$.

2.3.1. Degeneracy

This example also illustrates the idea of degeneracy of basic solutions. In finite-dimensional linear programs, degeneracy occurs when the basis has too few elements; that is, it is related to a shortfall in the dimension of $B(x)$. A degenerate basic solution does not define a unique complementary dual solution, a difficulty which is solved in the finite-dimensional simplex algorithm by arbitrarily adjoining further elements to the basis until a complementary dual solution is defined. The properties of the finite-dimensional linear program are such that this procedure causes no essential difficulty for the algorithm, and degeneracy in the finite-dimensional case is a relatively unimportant phenomenon. We shall see during the course of this book that degeneracy is far more important in the infinite-dimensional case.

For any $x \in P$, define $D(x) = B(x) + N(A)$. If x is a basic feasible solution for LP, $B(x) \cap N(A) = \{\theta\}$, so $D(x) = B(x) \oplus N(A)$. For any $x \in F$, $F \subset D(x)$. A basic feasible solution x is called *non-degenerate* if $D(x) = X$, *degenerate* otherwise. Thus a non-degenerate basic feasible solution has the property that for any z in AX, there is some $\xi \in B(x)$ with $A\xi = z$. Notice that in Perold's example, the basic solution ξ_0 is degenerate, since the constant function is not a member of $D(\xi_0)$. We shall see later that when LP has a non-degenerate basic optimal solution, the duality theory of finite LP carries over in its entirety to LP and AD. A corollary of this statement will be the fact that if a basic but possibly degenerate optimal solution exists, this duality relationship will hold between LP and AD if they are re-posed in $D(x)$ and its algebraic dual respectively. The proofs of these assertions begin with the extension to the current setting of the procedure for constructing complementary dual solutions. Before investigating this, we consider one more property of basic solutions, and look briefly at the question of whether or not basic solutions will exist in any particular case.

2.3.2. Minimal support

An alternative characterization of the basic solutions for finite-dimensional LP is that their support is minimal. That is, the statement that x is a basic feasible solution is equivalent to the statement that x is feasible, and there is no feasible ξ with $B(\xi)$ a proper subspace of $B(x)$. This property allows one to show easily that if an optimal solution exists, a basic optimal solution must exist. Given an optimal solution ξ_1 which is not basic, one simply constructs a new optimal solution ξ_2 for which $B(\xi_2)$ is a proper subspace of $B(\xi_1)$. If necessary, this process is repeated, and leads eventually to a basic optimal solution. To follow this route, we need some further assumptions. However the following weaker characterization of basic solutions is possible without any additional assumptions.

Theorem 2.3 *If x is a feasible solution, then the following statements are equivalent:*

(*a*) *x is a basic solution*
(*b*) *if ξ is any feasible solution with $B(\xi) \subset B(x)$ (not necessarily strictly), then $\xi = x$.*

Proof. (a) $\Rightarrow$ (b) Since $\xi \in B(\xi)$, $\xi \in B(x)$, and hence so is $x - \xi$. Since both x and ξ are feasible, $Ax = A\xi = b$, and $x - \xi \in N(A)$. Thus $x - \xi \in B(x) \cap N(A)$, and since x is basic this implies $x - \xi = \theta$.

(b) $\Rightarrow$ (a) Suppose that x is feasible but not basic. Choose η, not equal to θ, with $\eta \in B(x) \cap N(A)$, then there is some $\lambda \in R$ with $x + \lambda\eta \geqslant \theta$ and $x - \lambda\eta \geqslant \theta$. Let $\xi = x + \lambda\eta$, so that ξ is feasible and not equal to x. We show that $B(\xi)$ is contained in $B(x)$. If $\xi' \in B(\xi)$, then we may choose $\mu \in R$ with $x + \lambda\eta + \mu\xi' \geqslant \theta$ and $x + \lambda\eta - \mu\xi' \geqslant \theta$. Now

$$x + (\mu/2)\xi' = [(x - \lambda\eta) + (x + \lambda\eta + \mu\xi')]/2 \geqslant \theta$$

and similarly $x - (\mu/2)\xi' \geqslant \theta$. Hence $\xi' \in B(x)$. □

The rather weak characterization provided by this theorem is of little use. It can be strengthened if we make some further assumptions about the order structure; that is, some further assumptions about P. The order relation $\geqslant$ is called *Archimedean* if the statement $x + n\xi \geqslant \theta$, $n = 1, 2, \ldots$ necessarily implies $\xi \geqslant \theta$. This property of the order relation is essentially an algebraic closure property of P. To see this, note that if the order is Archimedean then $x/n + \xi \in P$, $n = 1, 2, \ldots$ implies $\xi \in P$. If L is a line in X and the order is Archimedean, then $L \cap P$ is a closed interval in L, for $L \cap P$ is convex, and hence consists of all of L, or is empty, or is an interval with either 1 or 2 endpoints. If ξ is such an endpoint, then for any point $x \in L \cap P$, $(x - \xi)/n + \xi \in P$, $n = 1, 2, \ldots$, and so $\xi \in P$.

If the order is Archimedean, and in addition $P \cap (-P) = \{\theta\}$, Theorem 2.3 can be sharpened.

Theorem 2.4 *If the order of X is Archimedean, and $P \cap (-P) = \{\theta\}$, then a feasible solution x is basic if and only if there is no other feasible solution ξ with $B(\xi)$ strictly contained in $B(x)$.*

Proof. We have proved already that if x is a basic feasible solution, there is no other feasible solution ξ with $B(\xi) \subset B(x)$. We now show that, under the assumptions of this theorem, if x is feasible but not basic, a feasible solution ξ exists with $B(\xi) \subset B(x)$, the inclusion being strict. So suppose that x is feasible, and that $x' \in B(x) \cap N(A)$. Let L be the line $\{x + \lambda x' \colon \lambda \in R\}$. Then $L \cap F = L \cap P$, and since P cannot contain the whole of any line by assumption, this intersection must be an interval in L which includes at least one end point. Let ξ be such an end point. Then $\xi \in B(x)$, and we have shown in the proof of theorem 2.3 that this means $B(\xi) \subset B(x)$; since it is clear that $B(\xi)$ cannot contain x', the inclusion is strict. □

2.3.3 Existence of a basic optimal solution

Consider the set $\{B(\xi): \xi \in F\}$ of subspaces of X, ordered by inclusion. Under the conditions of Theorem 2.4, a feasible solution x is basic if and only if $B(x)$ is minimal with respect to this ordering. In finite-dimensional LP, this fact is used to show that if a program has an optimal solution, then it has a basic optimal solution. Using the same idea, the following theorem provides a simple extension of this result.

Theorem 2.5 *Suppose that the conditions of Theorem* 2.4 *apply, and that* LP *has an optimal solution* x. *Suppose further that* $B(x) \cap N(A)$ *is of finite dimension. Then* LP *has a basic optimal solution.*

Proof. If x itself is basic, we are done. Otherwise, we can find $\xi \in B(x)$ feasible for LP, with $B(\xi) \cap N(A)$ strictly contained in $B(x) \cap N(A)$, by the construction of Theorem 2.4. Since $B(x) \cap N(A)$ is of finite dimension, repetition of this process if necessary will produce a feasible solution, still called ξ, which is basic and with $\xi \in B(x)$. We show that ξ is still optimal by showing that c^* annihilates $B(x) \cap N(A)$, and hence that $\langle (\xi - x), c^* \rangle = 0$. Suppose that $\zeta \in B(x) \cap N(A)$, and is not equal to θ. Then there exists $\lambda > 0$ such that $x + \lambda\zeta$ and $x - \lambda\zeta$ are both feasible. Clearly, if $\langle \zeta, c^* \rangle$ is other than zero, one of these must give a strictly improved objective functional value, contradicting the assumed optimality of x. □

Most orderings which arise in problems in practice are Archimedean. Examples of non-Archimedean orders are provided by sequence spaces under their lexicographic orderings. It is easy to find examples in such spaces where LP has an optimal solution, but no basic optimal solutions.

Example
Let $X = R^2$, and define P by

$$P = \{x\text{: either } x_1 > 0 \text{ or both } x_1 = 0 \text{ and } x_2 \geqslant 0\}.$$

Consider the problem

$$\begin{array}{ll} \text{minimize} & x_2 \\ \text{subject to} & x_2 \geqslant -1, \\ & x \geqslant \theta. \end{array}$$

The set of optimal solutions for this problem is $\{x: x_1 > 0, x_2 = -1\}$, and includes no extreme point of the feasible set.

Granted that P contains no line and defines an Archimedean order, the most obvious cases in which Theorem 2.5 applies are those where either $B(x)$ or $N(A)$ is of finite dimension. This is the situation, for example, for semi-infinite linear programs of the form

$$\begin{aligned} &\text{minimize} && c^{\mathrm{T}}x \\ &\text{subject to} && a(s)^{\mathrm{T}}x \leqslant b(s),\ s \in S, \\ & && x \in R^n, \quad x \geqslant 0, \end{aligned}$$

where S is an arbitrary index set. Here, after addition of slack variables $z(s)$, $s \in S$, we have

$$N(A) = \{(x, z) \colon z(s) = -a(s)^{\mathrm{T}}x, s \in S\}.$$

This subspace has dimension n whatever the size of S.

It is tempting to continue with this line of investigation, in an attempt to extend Theorem 2.5 further. To do that, we would have to seek conditions under which the set of subspaces $\{B(\xi)\colon \xi \in F\}$ is inductively ordered by inclusion, then apply Zorn's lemma to prove the existence of a ξ for which $B(\xi)$ is minimal. Thus we would try to guarantee that intersections of nested sequences of these subspaces are nonempty. There are some powerful theorems which provide the sort of conditions that are needed, but they make use of topological properties of X and F. We shall therefore leave further investigation of the question of existence of basic optimal solutions until the next chapter, where LP is posed in a setting which naturally endows the underlying spaces with a topological structure. We turn here to the question of how reduced costs can be constructed for a given basic solution.

2.4 REDUCED COSTS

We wish to construct an analogue of the reduced-cost vector of finite linear programming. We have already remarked that the reduced-cost vector can be characterized as a complementary dual solution and as a reduced gradient of the objective functional, and it is from this latter characterization that we start. Suppose that x is a basic feasible solution for LP. Let $M(x)$ be any direct complement of $B(x)$ in $D(x)$. Suppose $x + \xi$ is some other feasible solution. Then we can write

$$\xi = \xi_B + \xi_M,$$

with $\xi_B \in B(x)$ and $\xi_M \in M(x)$. Since $x + \xi$ is feasible, $\xi \in N(A)$, so that

$$A\xi_B = -A\xi_M.$$

Let A_B denote the restriction to $B(x)$ of A. Since x is basic, A_B is injective, hence invertible, so we may write

$$\xi_B = -A_B^{-1}A\xi_M.$$

The change in value of the objective functional can thus be computed as

$$\begin{aligned}\langle \xi, c^* \rangle &= \langle (\xi_B + \xi_M), c^* \rangle \\ &= \langle (I - A_B^{-1} A)\xi_M, c^* \rangle, \\ &= \langle \xi_M, (I - A_B^{-1} A)^* c^* \rangle.\end{aligned}$$

Here we have used I to represent the identity map. We define the *reduced cost functional* x^* by

$$x^* = (I - A_B^{-1} A)^* c^*.$$

Note that x^* is defined only on $D(x)$, not necessarily on all of X, and that it annihilates $B(x)$. The notation we have used above, particularly in the definition of x^*, has considerable intuitive content, but is not particularly convenient. It can be condensed by observing that $N(A)$ is itself a direct complement of $B(x)$ in $D(x)$, and that the map $(I - A_B^{-1} A)$ is just the projection of $D(x)$ onto $N(A)$. We denote this projection by P_N, and the corresponding projection onto $B(x)$ by P_B. Thus $x^* = P_N^* c^*$.

We now show that the reduced cost provides us with a check on the optimality of the basic solution x.

Theorem 2.6 *If x is a basic solution for* LP, *then x is optimal if and only if x^* is a positive functional on $D(x)$.*

Proof. We show first that if x^* is not positive, then x cannot be optimal. For if x^* is not positive, then there is some $\xi \in D(x)$ such that $\xi \geqslant \theta$ and $\langle \xi, x^* \rangle < 0$. Since $\xi \in D(x)$, we may set $\zeta = P_B(\xi)$; thus there is some $\lambda > 0$ such that $x - \lambda\zeta \geqslant \theta$, and hence

$$\begin{aligned}x - \lambda\zeta + \lambda\xi &\geqslant \theta, \\ A(x - \lambda\zeta + \lambda\xi) &= b.\end{aligned}$$

Moreover,

$$\begin{aligned}\langle x - \lambda\zeta + \lambda\xi, c^* \rangle &= \langle x, c^* \rangle + \lambda \langle \xi - \zeta, x^* \rangle, \\ &= \langle x, c^* \rangle + \lambda \langle \xi, x^* \rangle \\ &< \langle x, c^* \rangle.\end{aligned}$$

Hence x is not optimal.

Now suppose that x^* is positive. If ξ is any other feasible solution, we have

$$\begin{aligned}\langle x, c^* \rangle - \langle \xi, c^* \rangle &= \langle (x - \xi), c^* \rangle \\ &= \langle P_N(x - \xi), c^* \rangle \\ &= \langle x - \xi, x^* \rangle.\end{aligned}$$

Since x^* annihilates $B(x)$, then in particular $\langle x, x^* \rangle = 0$. Thus this last expression is just equal to $-\langle \xi, x^* \rangle$. Since ξ is feasible, it is positive, so $-\langle \xi, x^* \rangle \leqslant 0$. Hence

$$\langle x, c^* \rangle \leqslant \langle \xi, c^* \rangle,$$

and as ξ is an arbitrary feasible solution, x is optimal. □

We remark here that the assumption that x is basic is needed in the above proof only to guarantee that P_N is well defined and annihilates $B(x)$. Even if x is not basic, essentially the same arguments as those used above can be applied to the space $\langle x \rangle \oplus N(A)$, where $\langle x \rangle$ denotes the one-dimensional subspace spanned by x, provided that b is non-zero. It is easily verified that if we carry through the theory of this section with $D(x)$ replaced by this smaller space, we define a reduced cost which is just the restriction to $\langle x \rangle \oplus N(A)$ of the reduced cost functional x^* as defined here.

2.5 THE PIVOT STEP

Later in this chapter, we will show that the characterization of the reduced cost functional as a complementary solution to the dual constraints remains valid, the dual problem being re-posed in the algebraic dual space of $D(x)$. For the moment, we have shown that a reduced cost functional exists, and that its positivity serves as an optimality check. When it is not positive, feasible perturbations exist which produce an improvement in the objective functional value. This leads us to ask whether such a perturbation can be found which leads to an improved basic feasible solution.

Trivially, if an improved basic feasible solution exists, then some perturbation will take us to it. The question is rather one of characterizing these perturbations in a way which is of help in constructing analogues to the pivot step in the simplex algorithm. We begin by characterizing the set of all feasible perturbations from a given basic solution x. Define $K(x)$, the cone of feasible directions at x, by

$$K(x) = \{\xi \in X : x + \lambda\xi \in F \text{ for some } \lambda > 0\}.$$

Lemma 2.7 *If x is a basic feasible solution, then*

$$K(x) = P_N(D(x) \cap P).$$

Proof. Suppose first that $\xi \in D(x)$, $\xi \geqslant \theta$. We show that $P_N\xi \in K(x)$. We may write $\xi = P_N\xi + P_B\xi$. Since $P_B\xi \in B(x)$, there is some $\lambda > 0$ such that $x - \lambda P_B\xi \geqslant \theta$. Hence

$$x + \lambda P_N\xi = x + \lambda\xi - \lambda P_B\xi \geqslant \theta.$$

Since $A(x + \lambda P_N\xi) = Ax = b$, $x + \lambda P_N\xi$ is feasible.

Now suppose $\xi \in K(x)$. Then there exists $\lambda > 0$ such that $x + \lambda\xi$ is feasible. Clearly, this implies $\xi \in N(A)$. Now $\lambda^{-1}x \in B(x)$, thus $\xi = P_N(\lambda^{-1}x + \xi)$. Since $\lambda^{-1}x + \xi \in D(x)$, and is positive, $\xi \in P_N(D(x) \cap P)$. □

This characterization of $K(x)$ is implicit in the proof of Theorem 2.5: x is optimal if and only if $\langle \zeta, c^* \rangle \geqslant 0$ for all $\zeta \in K(x)$, that is, for all $P_N \xi, \xi \in D(x) \cap P$. This happens if and only if $P_N^* c^*$ is positive on $D(x)$. When x is not optimal, there exists some $\zeta \in K(x)$ such that $\langle \zeta, c^* \rangle < 0$, and hence some $\xi \in D(x) \cap P$ such that $\langle \xi, x^* \rangle < 0$. The problem in constructing a pivot step lies in recognizing those members of $D(x) \cap P$ for which this is true, and choosing from among them a direction which leads to another basic solution. Thus at this point one would like to produce a general theory of pivoting, extending to the infinite-dimensional case the finite-dimensional pivoting computation. What we find, however, is that the pivot step of the finite-dimensional simplex algorithm depends on certain properties of R^n and its order structure which are not necessarily found in a general space. Thus we can characterize the pivot step of finite-dimensional LP in abstract terms, but the result is useful for infinite-dimensional problems only in certain cases.

First, we remark that the idea of moving along a perturbation until we reach a boundary of the feasible set is itself sensible only if the order is Archimedean. If not, there may be perturbations ξ such that $\lambda = \sup\{\mu : x + \mu\xi \in F\}$ exists, but $x + \lambda\xi$ is not feasible. When the order is Archimedean, some of the perturbations which lead to new basic solutions will correspond, intuitively speaking, to 'edges' of F. That is, if ξ is such a perturbation, we expect the set $\{\zeta : \zeta \in F, \zeta = x + \mu\xi\}$ to be a one-dimensional extremal subset of F. The following theorem gives precise expression to that expectation.

Theorem 2.8 *Suppose x is a basic feasible solution for* LP, *and that $\{\mu\xi : \mu \geqslant 0\}$ is an extremal ray of $K(x)$, such that $\lambda = \sup\{\mu : x + \mu\xi \in F\} < \infty$ and $x + \lambda\xi \in F$. Then*

(a) $x + \lambda\xi$ is a basic feasible solution,
(b) if $0 < \mu < \lambda$, then $\dim(B(x + \mu\xi) \cap N(A)) = 1$.

Proof. (a) Suppose that $\zeta \in B(x + \lambda\xi) \cap N(A)$. We show first that $B(x + \lambda\xi) \subset B(x + \mu\xi)$ for any μ such that $0 < \mu < \lambda$. For if $\eta \in B(x + \lambda\xi)$ then there is some $\rho \in R$ with $x + \lambda\xi + \rho\eta \geqslant 0$. Now, setting $k = \mu/\lambda$,

$$x + \mu\xi + (k\rho)\eta = k(x + \lambda\xi + \rho\eta) + (1 - k)x \geqslant 0.$$

Similarly $x + \mu\xi - (k\rho)\eta \geqslant 0$, so that $\eta \in B(x + \mu\xi)$. Hence for any such μ, there is some ν such that $x + \mu\xi + \nu\zeta \in F$ and $x + \mu\xi - \nu\zeta \in F$. This contradicts the assumption that ξ defines an extremal ray of $K(x)$, unless ζ is proportional to ξ. However, the definition of λ ensures that ξ is not contained in $B(x + \lambda\xi)$. Hence $\zeta = \theta$ and $x + \lambda\xi$ is basic.

(b) Follows from a similar argument to the above. □

In the finite-dimensional simplex algorithm, the pivot operation corresponds to exactly this type of perturbation. Among all the edges of the feasible set originating at the current basic solution, one is chosen which produces an improvement, and the new solution is just the other end of that edge. In order to be able to carry out the same manoeuvre in an infinite-dimensional problem, we require that $K(x)$ have extremal rays, and that we should be able to characterize them. In addition, we must be able to tell which of them lead to an improved solution. In the finite-dimensional case, this last is simple, because we can decompose $D(x)$ into $B(x)$ and an *order direct complement* M (that is, if $\xi \in D(x)$, $\xi \geqslant \theta$, then $\xi = \xi_B + \xi_M$ and necessarily $\xi_B \geqslant \theta$ and $\xi_M \geqslant \theta$). The extreme rays of $K(x)$ are then just those vectors ξ for which $A\xi_B = -A\xi_M$ and ξ_M is an extremal ray of $M \cap P$. The extremal rays of $M \cap P$ have exactly one component non-zero, and among these it is straightforward to select one which has $\langle \xi_M, x^* \rangle$ negative.

In many infinite-dimensional problems, particularly those set in spaces of continuous functions, these characteristics are lacking. $K(x)$ may lack extremal rays, and $B(x)$ may have no order direct complement. Thus a general theory of pivoting which provides formulae analogous to those of finite-dimensional LP cannot be given. What we are able to do is state that in practice it is usually possible to find an appropriate pivot operation for a given infinite-dimensional LP, and to point to the variety of examples given in the later chapters of this book.

2.6 ALGEBRAIC DUALITY THEORY

We have not so far asserted the existence of an optimal solution to the dual problem AD with the same objective functional value as the optimal solution to the primal problem LP. In finite linear programming, there are two routes to a proof of this assertion. One is via a separation theorem applied to suitably constructed sets, the coefficients of the separating hyperplane yielding the required dual solution (see, for example, Luenberger (1973)). The other is via the existence of basic optimal solutions for the primal, and uses complementarity to construct a solution to the dual constraints. This solution is shown to be feasible, and hence optimal, by considering its interpretation as the reduced cost vector. This is straightforward when the given basic optimal solution is non-degenerate, but requires extra arguments, based on continuity of the optimum with variations in the constraint right-hand sides, when there is degeneracy.

In this section, we will examine the possibility of using this latter approach to prove that LP and AD are strongly dual. The key ideas are the algebraic characterization of the basic solutions and degeneracy. We begin by showing that the reduced cost functional can indeed be interpreted as a complementary dual solution. Suppose that x is a non-degenerate basic feasible solution for LP, and assume for the moment that $AX = Z$. As before, let A_B be the restriction of A to $B(x)$. Then A_B is a bijection from $B(x)$ to Z. Let $H = A_B^{-1}$ be its inverse, and define $z^* \in Z^*$ by $z^* = H^* c^*$.

Theorem 2.9 *If x is a non-degenerate basic feasible solution, x^* the associated reduced cost functional, and $AX = Z$, then (x^*, z^*) is the unique solution in $X^* \times Z^*$ of*

$$A^* z^* + x^* = c^*$$

which satisfies $\langle \xi, x^* \rangle = 0$ for all $\xi \in B(x)$.

Proof. We already know that x^* satisfies the complementarity condition. Observe directly from the definitions of x^* and z^* that $A^*z^* + x^* = c^*$. To show that (x^*, z^*) is unique, suppose that $A^*\zeta^* + \xi^* = c^*$, and ξ^* annihilates $B(x)$. Then $A^*\zeta^* = A^*z^* = c^*$ on $B(x)$, which implies that $\zeta^* = z^*$ on $AB(x)$. Since x is non-degenerate, $AB(x) = AX = Z$, and so $\zeta^* = z^*$ on all of Z. Hence $\xi^* = x^*$ as well. □

We show now that the existence of a non-degenerate basic optimal solution implies that LP and AD are strongly dual.

Theorem 2.10 *Suppose* $AX = Z$, *and that* x *is a non-degenerate basic solution for* LP. *Then* x *is optimal if and only if its complementary dual solution* (x^*, z^*) *is optimal for* AD.

Proof. If the complementary dual solution is feasible, then x is optimal for LP by weak duality. On the other hand, if x is optimal and non-degenerate, x^* is a positive functional on X by Theorem 2.6. Thus (x^*, z^*) is feasible for AD, and hence optimal by weak duality. □

The assumption that $AX = Z$ is unnecessary in the last two theorems, and has been made solely to avoid notational complexity. If it does not hold, we can just re-pose LP with its constraints in $Z_1 = AX$, and use Theorem 2.10 to guarantee the existence of (x^*, z^*) with $z^* \in Z_1^*$ which solves the appropriate version of AD. We can then extend z^* to all of Z in any way we choose. The form of the constraint in AD means that the value of z^* outside Z_1 is irrelevant.

It is tempting to suggest that the same procedure could be used to remove the restriction that x should be a non-degenerate basic optimal solution. The idea would be to repose LP on $D(x)$, obtain a solution (x^*, z^*) to the new dual, and then extend x^* to the whole of X by setting it equal to zero on a direct complement of $D(x)$. The snag, of course, is that this extension would have to ensure that x^* remained positive.

In the finite-dimensional problem, this difficulty is very easily circumvented. There $B(x)$ is just the subspace $\{\xi : \xi_i = 0 \text{ if } x_i = 0\}$; x is non-degenerate if m of its components are non-zero, m being the rank of A. If fewer than m components of x are positive, $B(x)$ is extended by arbitrarily treating as positive sufficient of the zero components of x. Whether the resulting extension of x^* is positive depends on the choice of components introduced into the basis. However, there is only a finite number of such choices, and perturbation arguments can be used to show that there is at least one such extension such that the constructions described above lead to an optimal dual solution, complementary with x. Unfortunately, there are several reasons why this will not work in general. First, it depends on the detailed structure of R^n under its natural order. Secondly, the stability properties which make the perturbation arguments work are absent in general. Finally, even if the method can be adapted to a particular infinite-dimensional problem, there may well be an infinite number of possible extensions, only some of which will be positive.

The question of when a positive linear functional on a subspace can be extended to a positive linear functional on the whole space can be answered very precisely. The theorem of Bauer and Namioka (see, for example, Schaefer (1971) 5.4) states that a positive linear functional x^* on some subspace M of a space X has a positive extension to X if and only if x^* is bounded above on $M \cap (U-P)$, for some convex, radial subset U of X which contains the origin (a subset U is *radial* if for each $\xi \in X$ there exists $\lambda \in R$ such that $\xi \in \lambda U$). The Krein–Rutman theorem (Holmes (1975) 6B) provides a simple sufficient condition: if M contains an internal point of P, then every positive linear form on M has a positive extension to X (a point ξ is *internal* in a set P if there is some radial set U with $\theta \in U$ such that $\xi + U \subset P$). The following theorem is a slight extension of the Krein–Rutman theorem.

Theorem 2.11 *Suppose that M is a subspace of X, and that for each $x \in X$ there exists $\xi \in M$ such that $x \leqslant \xi$. Then every positive linear functional on M has a positive extension to X.*

Proof. We construct a suitable radial set U and apply the theorem of Bauer and Namioka. Suppose that x^* is a positive linear functional on M, and define $V = \{\xi\colon\ \xi \in M,\ \langle \xi, x^* \rangle < 1\}$. Then define $U = \{x\colon x \leqslant \xi \text{ for some } \xi \in V\}$. It is not hard to check that U is a convex radial set containing the origin. If $x \in M \cap (U-P)$, then there is a $u \in U$ with $x \leqslant u$, and hence $x \leqslant \xi$ for some $\xi \in V$. Thus if $\zeta = \xi - x$, then $\zeta \geqslant \theta$. Moreover $\zeta \in M$, and so

$$\langle x, x^* \rangle = \langle \xi, x^* \rangle - \langle \zeta, x^* \rangle < 1,$$

as x^* is positive on M. □

We define the *P-saturated hull* of a subset S of X, denoted $[S]$, by $[S] = (S-P) \cap (S+P)$. Theorem 2.11 allows us to say that a positive linear functional on a subspace M always has a positive extension to $[M]$. Recalling our earlier remark that all of Section 2.4 can be carried through with $D(x)$ replaced by $\langle x \rangle \oplus N(A)$, we note that $B(x) = [\langle x \rangle]$. This follows from observing that if $x + \lambda w \geqslant \theta$ and $x - \lambda w \geqslant \theta$ with $\lambda > 0$, then $(x/\lambda) \geqslant w \geqslant -(x/\lambda)$, and if $\lambda_1 x \geqslant w \geqslant \lambda_2 x$, then setting $\mu^{-1} = \max\{|\lambda_1|, |\lambda_2|\}$ makes $x + \mu w \geqslant \theta$ and $x - \mu w \geqslant \theta$. Hence if x^* is a reduced cost defined on $\langle x \rangle \oplus N(A)$ for an optimal solution x, then x^* extends to a positive functional on $D(x)$. It is not hard to see that since x^* annihilates x, it must annihilate $B(x)$. This argument does not rely on x being basic, and so we have shown that for any optimal solution x of LP, there is a reduced cost functional defined on $D(x)$ which is positive. Of course, only if x is basic can the reduced cost be constructed (as a functional on $D(x)$) by the formulae of Section 2.4, while the interpretation of x^* as a complementary dual solution and its associated uniqueness properties similarly depend on x being basic.

Thus far, then, we have derived several conditions sufficient to guarantee strong duality of LP and AD once a basic optimal solution exists for LP. These are that x be non-degenerate, that the conditions of the Bauer–Namioka or Krein–Rutman theorems hold for $D(x)$, or that every element of X be majorized by some element of $D(x)$. Where none of these hold, we can still guarantee strong duality if LP is reposed in $[D(x)]$ and AD re-posed in appropriate dual spaces. In the theory of optimization in general, there is a class of conditions (Slater conditions) which are used to guarantee strong duality results. We end this section by showing that some of our conditions imply other conditions of Slater type.

Theorem 2.12 *If $AX = Z$, and either*

(*a*) LP *has a non-degenerate basic feasible solution, or*
(*b*) $D(x)$ *contains an internal point of P, for some feasible solution x,*

then b is internal in AP.

Proof. (a) Suppose x is a non-degenerate basic feasible solution. Then $AB(x) = Z$, and the order interval $[-x, x] = \{\xi\colon -x \leqslant \xi \leqslant x\}$ is radial in $B(x)$. Thus $A[-x, x]$ is radial in Z. Since $x + [-x, x] \subset P$, $A(x + [-x, x]) = b + A[-x, x] \subset AP$.

(b) Suppose $\xi \in D(x)$ is an internal point of P, and write $\xi = \xi_1 + \xi_2$, where $\xi_1 \in B(x)$ and $\xi_2 \in N(A)$. Then for some $\lambda > 0$, $x - \lambda\xi_1 \geqslant \theta$, so $x = \lambda\xi_1 + \xi_3$, for some $\xi_3 \in P$. Suppose that U is a radial set in X such that $\xi + U \subset P$. Then λU is also radial in X, and $A(\lambda U)$ is radial in Z. We show that $b + A(\lambda U) \subset AP$.

$$
\begin{aligned}
b + A(\lambda U) &= A(x + \lambda U) \\
&= A(\lambda U + \lambda\xi - \lambda\xi_2 + \xi_3) \\
&= A(\lambda U + \lambda\xi + \xi_3) \subset AP.
\end{aligned}
$$
□

2.7 NOTES

It is perhaps surprising that very little work has been published which deals with the questions raised in this chapter. It seems that the only paper with much in common with the approach taken here is that by Meidan and Perold (1983). They work with $\langle x \rangle \oplus N(A)$ rather than $D(x)$, and do not address the properties of basic solutions. Meidan and Perold work in spaces with an explicit topological structure, although they make use of this only when they consider the extension of positive functionals on $\langle x \rangle \oplus N(A)$ to X, for which they use topological versions of the Bauer–Namioka and Krein–Rutman theorems.

Meidan and Perold also relate the boundedness condition which results from the Bauer–Namioka theorem to a stability condition of Rockafellar (1967). This condition is that the *optimal value functional*,

$$\omega(z) = \inf\{\langle x, c^* \rangle\colon Ax = z, x \geqslant \theta\}, z \in Z,$$

have bounded directional derivatives at b. More precisely the condition is that there be a neighbourhood of the origin $U \subset Z$ with

$$\lim_{\varepsilon \to 0} (\omega(b) - \omega(b + \varepsilon z))/\varepsilon$$

bounded above for each $z \in U$.

In looking for extensions of the simplex algorithm, the results of this chapter are sufficient for practical purposes. The value of some of the stronger duality results that can be obtained within a topological framework, and which we shall examine in detail in the next chapter, stems from the fact that the algebraic duals of most infinite-dimensional spaces do not admit of any simple concrete characterization. From a theoretical point of view, it is thus of considerable interest to see when the dual problem can be posed in a space whose elements do have a useful characterization. The most powerful tools for investigating this question are those of functional analysis, where we consider subspaces of the algebraic dual space whose members are continuous in some topology on the primal space. Given such a topology, we can impose conditions on LP such as to guarantee that x^* is continuous in the canonical topology induced on $D(x)$, then use topological versions of the Bauer–Namioka theorem to provide conditions under which x^* can be extended to all of X while retaining positivity and continuity. As we remarked at the end of Section 2.4, working with the tools of functional analysis in a topological framework is also the best way to attack the question of existence of extreme point solutions for LP.

A number of authors have been concerned to set up linear programs in a more general algebraic setting that we have considered here. One of the motivations for this is a desire to encompass certain combinatorial problems not usually thought of as linear programs. An extensive treatment of this area can be found in Zimmermann (1981). In this field, most attention has been given to the duality theory of linear programs posed over totally ordered semimodules (see, for example, Burkard, Hamacher and Tind (1982)).

Problems

2.1 Consider the following finite linear program:

$$\begin{aligned} &\text{minimize} && x_1 - 2x_2 \\ &\text{subject to} && 2x_1 + x_2 + x_3 = 1, \\ & && -2x_1 + 3x_2 + x_4 = 1, \\ & && x_i \geqslant 0, \qquad i = 1, 2, 3, 4. \end{aligned}$$

Show that $x = (\frac{1}{4}, \frac{1}{2}, 0, 0)$ is a basic feasible solution which is non-degenerate. Find P_B and P_N for this solution and hence show that the reduced-cost functional, x^*, is non-negative.

2.2 We have chosen to work with vector spaces over the real field. Though convenient, this is not essential. For example, suppose that X and Z are vector spaces over the complex numbers, C, and A is a linear map from X to Z. Let X^*

and Z^* be the spaces of all linear maps from X and Z respectively to C (so for x^* in X^*, $\langle x, x^* \rangle$ is complex). Let $C^+ = \{z \in C: |\arg z| \leqslant \alpha\}$ where α is some given constant with $0 \leqslant \alpha \leqslant \pi/2$. Now for any cone S, in X say, we can define its dual cone S^* by

$$S^* = \{x^* \in X^*: \langle s, x^* \rangle \in C^+ \quad \text{for all} \quad s \in S\}.$$

With these definitions we define the primal linear program as follows:

$$\begin{array}{ll} \text{minimize} & \operatorname{Re} \langle x, c^* \rangle \\ \text{subject to} & Ax = b, \qquad x \in P. \end{array}$$

This has associated with it a dual problem:

$$\begin{array}{ll} \text{maximize} & \operatorname{Re} \langle b, z^* \rangle \\ \text{subject to} & c^* - A^*z^* \in P^*, \qquad z^* \in Z^*. \end{array}$$

For this pair of problems prove a weak duality result similar to that of Theorem 2.1. (For more results on linear programming in a complex space, see Levinson (1966b)).

2.3 Consider the following problem:

$$\begin{array}{ll} \text{maximize} & \displaystyle\int_0^1 (1-t)\,x(t)\,\mathrm{d}t \\ \text{subject to} & \displaystyle\int_0^t x(s)\,\mathrm{d}s \leqslant 3t^2, \\ & 0 \leqslant x(t) \leqslant 3, \qquad t \in (0, 1), \end{array}$$

for $x(t)$ a continuous function. Show that

$$\begin{aligned} x(t) &= 6t, \qquad t \in (0, 1/2], \\ &= 3, \qquad t \in (\tfrac{1}{2}, 1), \end{aligned}$$

is a degenerate basic feasible solution.

2.4 Let x be feasible for LP and suppose that this problem has an optimal value V. Show that if there is $z^* \in Z^*$ and $d^* \in X^*$ with

$$\langle x, c^* - A^*z^* \rangle = 0,$$
$$A^*z^* \leqslant c^* + d^*,$$

and $\langle \xi, d^* \rangle \leqslant M$ for every ξ feasible for LP, then

$$V \leqslant \langle x, c^* \rangle \leqslant V + M.$$

2.5 Show that z^* is a basic solution for AD if and only if

$$N(A^*) = A^*Z^* \cap B(c^* - A^*z^*) = \{\theta\}.$$

2.6 A vector lattice is an ordered vector space in which supremums and infimums exist (see for example Schaefer (1980)). In such a space we can define

$|x| = \sup(x, -x)$. A subspace A of a vector lattice is called *solid* if $x \in A$ and $|y| \leqslant |x|$ implies $y \in A$. Show that if X is a vector lattice, $B(x)$ is the smallest solid subspace containing x.

2.7 (Probabilistic programming)
Let Y be a random variable defined on $[0, 1]$ with density function $f(y) = 2y$, $y \in [0, 1]$; it is required to find the distribution of a random variable X defined on $[0, 1]$ having the property that the probability that X is greater than Y is α, say, and minimizing the expectation of X. Show how this can be formulated as a linear program and find its optimal solution.

2.8 (Silent duel)
Consider the silent duel of Section 1.2.5, when a shot fired at time t has a probability t of killing the opponent. Show that the optimal solution is for the duellist to fire his shot at a random time t with probability density given by

$$f(t) = 0, \qquad 0 \leqslant t \leqslant \tfrac{1}{3},$$

$$f(t) = 1/(4t^3), \quad \tfrac{1}{3} \leqslant t \leqslant 1.$$

2.9 (L_1 approximation)
Consider the following L_1 approximation problem on $[0, 1]$: given analytic functions $f, g_1, g_2, \ldots, g_n$, find $\lambda = (\lambda_1, \lambda_2, \ldots, \lambda_n)$ to

$$\text{minimize} \int_0^1 |f(s) - \Sigma\lambda_i g_i(s)|\,ds.$$

Show how this can be formulated as a linear program. For a given solution λ, define the error function e by $e(s) = f(s) - \Sigma\lambda_i g_i(s)$. Let $s_1, s_2, \ldots, s_m$ be the zeros of e and suppose that $e'(s_i) \neq 0$, $i = 1, 2, \ldots, m$. Show that the given solution is basic if and only if the vectors $h_i = (g_i(s_1), g_i(s_2), \ldots, g_i(s_m))^{\mathrm{T}}$, $i = 1, 2, \ldots, m$, are linearly independent.

3

Topology and Duality

3.1 INTRODUCTION

Thus far, we have confined ourselves to a purely algebraic framework, using as little *a priori* structure as possible. Consequently the results obtained in the previous chapter are very general, and apply to almost any linear program in an ordered vector space. Part of the price of this degree of generality is that we have to pose our dual problem in the algebraic dual of the primal space, and this is usually a difficult space in which to work. In addition most of the results of the previous chapter assume the existence of basic solutions, often of basic optimal solutions. It is therefore natural at this point to seek conditions under which basic solutions exist, and conditions under which a dual problem can sensibly be posed in a smaller, better-behaved space.

The most powerful tools at our disposal in such an investigation are those of functional analysis; to use these tools we must endow the spaces in which we work with a topological structure. Once this is done, we will be able to apply the powerful results which are available concerning the existence of extreme points, and the topological versions of the extension and separation theorems, which will allow us to deduce conditions for the existence of dual solutions which are linear functionals continuous in some topology. By this means we can hope to solve our dual problems in spaces with usable, concrete characterizations.

In fact, existence of extreme points is a relatively uncomplicated question. As we have already remarked, one requires a condition under which the set of extremal manifolds of the feasible region is inductively ordered by inclusion, so that extreme points can be characterized as minimal elements of this set. In turn this requires that intersections of descending chains of extremal manifolds should be non-empty, and this can be guaranteed by a compactness property. Duality theory is much more complex. There is the possibility that strong duality may not hold, but one of a number of other, weaker forms of duality may be proved. Thus the possible duality states that can exist between an arbitrary primal – dual pair of problems are quite large in number.

Most of this chapter is concerned with categorizing these different duality states, and giving conditions under which one or other will hold. The relationship between primal and dual in such a discussion is more or less symmetric, and whereas we have previously taken the primal problem as a given starting point from which the dual is investigated, our formulation here strongly reflects this symmetry. We pose the problem in a way which uses dual pairs of vector spaces, rather than a topological vector space and its (topological) dual. This gives a formulation which is one of the most general possible, though in some places we impose restrictions in order to obtain stronger results. One penalty of this approach is that we shall require some results from the theory of dual pairs of vector spaces with which the reader may not be familiar. We begin by giving the results that we need.

3.2 DUAL PAIRS OF TOPOLOGICAL VECTOR SPACES

In this section we give a short review of the mathematical background we require. A fuller treatment can be found, for example, in Robertson and Robertson (1973) or in Schaefer (1971).

A *topological vector space* X is a vector space over R with a translation-invariant topology having a θ-neighbourhood base β with the properties that for each V in β:

(1) there is some $U \in \beta$ with $U + U \subset V$,
(2) for each $x \in X$ there is some $\lambda \in R$ with $x \in \lambda V$,
(3) for each $x \in V$, λx is in V for $-1 \leqslant \lambda \leqslant 1$.

A set V having property (3) is called *balanced*.

A topological vector space is *locally convex* if it has a base of convex sets. A locally convex topological vector space which is Hausdorff will be called a *locally convex space*.

Let X and Y be any two vector spaces with a bilinear form defined on $X \times Y$, i.e. some function from $X \times Y$ to R, which we write as $\langle \cdot , \cdot \rangle$, with $\langle x, y \rangle$ a linear function of x for each fixed y in Y, and a linear function of y for each fixed x in X. If

(1) for each $x \neq \theta$ in X there is some $y \in Y$ with $\langle x, y \rangle \neq 0$, and
(2) for each $y \neq \theta$ in Y there is some $x \in X$ with $\langle x, y \rangle \neq 0$,

then the pair of spaces (X, Y) is called a *dual pair*. Clearly (Y, X) is also a dual pair. If (X, Y) is a dual pair, then there is a natural injection of Y into the algebraic dual of X, obtained by identifying y with the linear functional $\langle \cdot , y \rangle$.

We can define a topology on X, called the *weak topology* and denoted by $\sigma(X, Y)$, by letting a base of neighbourhoods at the origin be the sets

$$B_A = \{x \in X\colon -1 \leqslant \langle x, y \rangle \leqslant 1 \quad \text{for all } y \in A\},$$

where A runs through all the finite subsets of Y. With this topology X is a locally convex space. The topology $\sigma(X, Y)$ is the coarsest topology on X under which all the elements of Y are continuous when regarded as linear forms on X. The result below states that the elements of Y regarded in this way are the only linear forms on X which are continuous with respect to this topology.

Proposition 1 *If (X,Y) is a dual pair of vector spaces, then Y is the dual of X with the topology $\sigma(X,Y)$.*

A locally convex topology on X is called *consistent with the dual pair* (X,Y) if Y is the dual of X with this topology. The coarsest such topology is $\sigma(X,Y)$. In other words it is the locally convex topology on X with the fewest open sets amongst those consistent with the dual pair (X,Y). There is also a finest such topology, called the *Mackey topology* and written $\tau(X,Y)$.

Proposition 2 *If (X,Y) is a dual pair of vector spaces and B is a convex set in X, then the closure of B is the same for every topology consistent with the dual pair (X,Y).*

Let (X,Y) be a dual pair of vector spaces. If B is a subset of X, then the *polar* of B is the set

$$B^0 = \{y \in Y: \quad \langle x, y \rangle \leqslant 1 \quad \text{for each } x \in B\}.$$

When B is a cone then its polar is evidently the cone $\{y \in Y: \langle x, y \rangle \leqslant 0$ for all $x \in B\}$.

Proposition 3 *If (X,Y) is a dual pair of vector spaces and B is a subset of X, then B^{00} is the $\sigma(X,Y)$-closed, convex hull of $B \cup \{\theta\}$.*

Suppose that (X,Y) and (Z, W) are two dual pairs of vector spaces; as there will be no danger of confusion, the bilinear forms on X and Y and on Z and W are both represented by $\langle \cdot, \cdot \rangle$. If A is a linear map from X to Z, then the adjoint (or transpose) of A, written A^*, is defined by the relation $\langle Ax, w \rangle = \langle x, A^*w \rangle$ for each x in X and w in W. The adjoint of A is a linear map from W to the algebraic dual of X, X^*. It is the restriction to W of the algebraic adjoint map, introduced in Chapter 2 and also called A^*, which is defined on Z^*.

Proposition 4 *A^* maps W into Y if and only if A is continuous with respect to the topologies $\sigma(X,Y)$ and $\sigma(Z, W)$.*

Proposition 5 *If A is a linear map from X to Z which is continuous with respect to some topologies on X and Z, which are consistent with the dual pairs (X,Y), and (Z, W), then A is continuous with respect to the topologies $\sigma(X,Y)$ and $\sigma(Z, W)$.*

We will need to use an analogue in a topological vector space setting of the separating hyperplane theorem of finite-dimensional convex analysis. There are many roughly equivalent ways of stating such theorems. The following version is the most convenient for our purposes.

Proposition 6 *If (X,Y) is a dual pair of vector spaces, B is a $\sigma(X,Y)$-closed convex set in X and a is an element of X not in B, then there is a y in Y and scalar c such that*

$\langle a, y\rangle < c$ and $\langle x, y\rangle > c$ for each x in B. (In these circumstances we say that a is strictly separated from B.)

3.3 THE LINEAR PROGRAMMING PROBLEM

Let (X,Y) and (Z, W) be two dual pairs of vector spaces. We adopt the convention that, wherever a topology on any space is implied, then it is to be taken as $\sigma(X,Y)$ or $\sigma(Z, W)$ as appropriate unless some other topology is used explicitly. Let P and Q be the positive cones in X and Z respectively, and let A be a $\sigma(X,Y)-\sigma(Z, W)$-continuous map from X to Z. Then we define an equality-constrained linear program, now called EP to distinguish it from the inequality-constrained program, as

$$\begin{aligned} \text{EP: minimize} \quad & \langle x, c\rangle \\ \text{subject to} \quad & Ax = b, \\ & x \geqslant \theta, \end{aligned}$$

where b and c are given elements of Z and Y respectively. The inequality constrained form of the linear programming problem (IP) is

$$\begin{aligned} \text{IP: minimize} \quad & \langle x, c\rangle \\ \text{subject to} \quad & Ax - b > \theta, \\ & x \geqslant \theta. \end{aligned}$$

These programs are formally the same as the linear programs introduced in Chapter 2. Recall that a program is called *consistent* if it has a feasible solution. The *value* of a consistent program EP (or IP) is defined as the infimum over feasible x of $\langle x, c\rangle$.

3.3.1 Elementary duality theory

In this section we shall introduce dual problems for EP and IP and make some elementary remarks about their properties. Almost everything we say can be carried over directly from the equivalent results for the finite linear program.

The *dual cones* of P and Q are

$$P^* = \{y \in Y\colon \langle x, y\rangle \geqslant 0 \text{ for all } x \text{ in } P\},$$

$$Q^* = \{w \in W\colon \langle z, w\rangle \geqslant 0 \text{ for all } z \text{ in } Q\}.$$

We have already used the notation P^* for the algebraic dual cone defined in Section 2.2. The dual cones we now define are the restrictions of the algebraic dual cones, defined on X^* and Z^*, to Y and W. As P and Q are cones, P^* and Q^* are the negative polars of P and Q.

The dual problems for EP and IP, called *EP** and IP* respectively, are defined as follows:

$$\begin{array}{lll} \text{EP*:} & \text{maximize} & \langle b, w \rangle \\ & \text{subject to} & -A^*w + c \in P^*, \\ & & w \in W, \end{array}$$

and

$$\begin{array}{lll} \text{IP*:} & \text{maximize} & \langle b, w \rangle \\ & \text{subject to} & -A^*w + c \in P^*, \\ & & w \in Q^*. \end{array}$$

Note that both these problems can be put in the form IP. This is done by making the following series of substitutions: $W \to X$, $Z \to Y$, $Y \to Z$, $A^* \to -A$, $b \to -c$, $c \to -b$, $P^* \to Q$ and W (or Q^*) $\to P$ and then noting the equivalence of maximizing $\langle x, -c \rangle$ to minimizing $\langle x, c \rangle$. If EP* (or IP*) is consistent its *value* is defined as the supremum over feasible w of $\langle b, w \rangle$.

We shall often refer to X as the *primal variable space*, Z as the *primal constraint space*, W as the *dual variable space* and Y as the *dual constraint space*. We can represent this schematically as follows:

$$\text{EP:} \begin{bmatrix} X \\ Y \end{bmatrix} \begin{matrix} \to \\ \leftarrow \end{matrix} \begin{bmatrix} Z \\ W \end{bmatrix} \text{:EP*}$$

The weak duality and complementary slackness theorems for EP and EP* follow from Theorem 2.1 and the canonical embeddings of Y and W in X^* and Z^*.

Theorem 3.1 (*Weak duality*) *If* EP *and* EP* *are both consistent, then the value of* EP *is greater than or equal to the value of* EP* *and both values are finite.*

Theorem 3.2 (*Complementary slackness*) *If x is feasible for* EP, *w is feasible for* EP* *and*

$$\langle x, c - A^*w \rangle = 0, \tag{1}$$

then x is optimal for EP *and w is optimal for* EP*.

There are corresponding results to Theorems 3.1 and 3.2 for the inequality constrained problem and its dual. However, in the statement of Theorem 3.2 the condition

$$\langle Ax - b, w \rangle = 0 \tag{2}$$

has to be added to condition (1). Conditions (1) and (2) are the complementary slackness conditions which we have met already in Chapter 2.

As EP* is a linear program it too has a dual program, called EP**. After introduction of slack variables this is easily seen to have the following form

$$\begin{aligned} \text{EP}^{**}: \quad \text{minimize} \quad & \langle x, c \rangle \\ \text{subject to} \quad & Ax = b, \\ & x \in P^{**}. \end{aligned}$$

Here P^{**} is the dual cone of P^*. Now, from the result given earlier on the polar of the polar of a set, it follows that P^{**} is the smallest closed convex cone containing P, and thus is equal to P if P is closed. So we can make the following important observation: if the positive cone P is closed, then EP and EP** are identical. The fact that a linear program is its own double dual if the positive cone is closed will allow us to write down dual versions as corollaries for many of the theorems which we give in the following sections.

3.4 SUBCONSISTENCY AND SUBVALUES

For a finite linear program the values of the primal and dual programs are always equal. This is not the case for infinite-dimensional linear programs. In order to understand what happens for these infinite linear programs we need to introduce the idea of a subvalue for a program.

Let D, H and L be sets in $Z \times R$ defined as follows:

$$D = \{(Ax, \langle x, c \rangle): x \in P\},$$

$$H = \{(Ax, \langle x, c \rangle + r): x \in P, r \geqslant 0\},$$

$$L = \{(b, r): r \in R\}.$$

The set H is obtained by extending D 'upwards', while L is a 'vertical' line through the point $(b, 0)$. It is clear that L has a non-empty intersection with D exactly when EP is consistent and the value of EP is just the infimum of r for which (b, r) is in $L \cap D$.

We now introduce three related conditions on the program EP.

Condition 1. b is in the closure of AP.
Condition 2. There is some r in R with (b, r) in the closure of H.
Condition 3. There is some r in R with (b, r) in the closure of D.

We shall write $\bar{D}$ and $\bar{H}$ for the closures of D and H in the $\sigma(Z \times R, W \times R)$ topology.

It is not difficult to see that Condition 3 implies Condition 2, which in turn implies Condition 1. We shall call the program EP *subconsistent* if Condition 2 holds. Equivalently EP is subconsistent if L has a non-empty intersection with $\bar{H}$. If EP is consistent it is certainly subconsistent but the reverse is not true. All three conditions above have been used to define subconsistency by other authors.

If Z has a countable base of neighbourhoods at the origin, then convergent sequences can be used to define the closure of sets in Z. This is the case for example if Z is a normed space. In this situation EP is subconsistent if there exists a sequence $\{x_n\}$ in P with $Ax_n \to b$ and $\langle x_n, c\rangle$ bounded above. It is often convenient to think of subconsistency as defined by the existence of sequences which approach feasibility in this sense.

If EP is subconsistent, its *subvalue* is defined as the infimum of r for which (b, r) is in $\bar{H}$. Thus, when we can interpret closure by convergent sequences, the subvalue is the infimum of r for which there exists a sequence $\{x_n\}$ in P with $Ax_n \to b$ and $\langle x_n, c\rangle \to r$. In these circumstances the slightly different definitions of subconsistency and subvalue used by some authors become identical.

The next theorem establishes the central result of this chapter: that there is never a duality gap between the primal subvalue and the dual value.

Theorem 3.3 EP *is subconsistent with a finite subvalue M if and only if* EP* *is consistent with a finite value M.*

Proof. The theorem is proved in two parts using an argument based on separating hyperplanes. We note first that H is a convex set, so we may apply Proposition 6.

(a) Suppose that EP is subconsistent with subvalue M. Then for $\varepsilon > 0$, $(b, M-\varepsilon)$ can be strictly separated from $\bar{H}$. Thus there is some $w \in W$ and $\lambda \in R$ with

$$\langle b, w\rangle + \lambda(M-\varepsilon) < \langle Ax, w\rangle + \lambda\langle x, c\rangle + \lambda r, \tag{3}$$

for every $x \in P$ and $r \geqslant 0$. Now suppose that $\langle x, A^*w + \lambda c\rangle$ is negative for some x in P. Then for α sufficiently large and positive,

$$\langle \alpha x, A^*w + \lambda c\rangle < \langle b, w\rangle + \lambda(M-\varepsilon). \tag{4}$$

However, the left-hand side of (4) is just $\langle A(\alpha x), w\rangle + \lambda\langle \alpha x, c\rangle$. As $\alpha x \in P$, this contradicts (3) with $r = 0$. Hence $A^*w + \lambda c \in P^*$.

Now suppose that $\lambda \leqslant 0$. Then

$$\langle b, w\rangle + \lambda M \leqslant \langle b, w\rangle + \lambda(M-\varepsilon),$$

and the hyperplane defined by (w, λ) also separates (b, M) from $\bar{H}$. This contradicts the definition of M, so $\lambda > 0$. Thus

$$-A^*(-(1/\lambda)w) + c \in P^*.$$

So $-(1/\lambda)w$ is feasible for EP*, and EP* is consistent.

Taking $x = 0$ and $r = 0$ in (3) we have

$$\langle b, w\rangle + \lambda(M-\varepsilon) < 0,$$

and so

$$\langle b, -(1/\lambda)w\rangle > M - \varepsilon.$$

Here ε was arbitrary, hence the value of EP* is at least M.

Now suppose that the value of EP* is greater than M. Then there is some $w_0 \in W$ with $\langle b, w_0\rangle$ strictly greater than M and $-A^*w_0 + c \in P^*$. Thus for every $x \in P$ and $r \geqslant 0$,

$$\langle b, -w_0\rangle + M < 0 \leqslant \langle x, -A^*w_0 + c\rangle + r.$$

Rewriting the right-hand side of this inequality in the form of (3), we see that (b, M) is strictly separated from H by the hyperplane defined by $(-w_0, 1)$. This contradicts the definition of M and so the value of EP* is exactly M.

(b) Suppose that EP* is consistent with value M. Then for every positive ε there exists some $w \in W$ such that $-A^*w + c \in P^*$ and $\langle b, w \rangle > M - \varepsilon$. Hence for every $x \in P$ and $r \geqslant 0$,

$$\langle Ax, -w \rangle + \langle x, c \rangle + r \geqslant 0 > \langle b, -w \rangle + M - \varepsilon.$$

Thus $(b, M - \varepsilon)$ is separated from H by a hyperplane defined by $(-w, 1)$, and so $(b, M - \varepsilon)$ is not in $\bar{H}$.

Now suppose that (b, M) is not in $\bar{H}$. Then there is some $w \in W$ and $\lambda \in R$ with

$$\langle b, w \rangle + \lambda M < \langle Ax, w \rangle + \lambda \langle x, c \rangle + \lambda r \tag{5}$$

for every $x \in P$ and $r \geqslant 0$. Notice that

$$\langle Ax, w \rangle + \lambda \langle x, c \rangle + \lambda r \geqslant 0, \tag{6}$$

for every $x \in P$ and $r \geqslant 0$. Otherwise we get a contradiction from (5) by multiplying x and r by α, for α sufficiently large and positive. Hence, putting $r = 0$, we have $A^*w + \lambda c \in P^*$. Setting $x = \theta$ in (6) gives $\lambda \geqslant 0$.

Now consider two cases depending on the value of λ. (i) If $\lambda > 0$, then $-(1/\lambda)w$ is feasible for EP*. Putting $x = \theta$ and $r = 0$ in (5) we have $\langle b, w \rangle + \lambda M < 0$, and so $\langle b, -(1/\lambda)w \rangle > M$. This contradicts the definition of M. (ii) If $\lambda = 0$ then let w_0 be feasible for EP*. As A^*w is in P^*, $w_0 - \alpha w$ is feasible for EP* for every $\alpha \geqslant 0$. Because EP* has a finite value, this implies that $\langle b, w \rangle$ is non-negative, which contradicts (5) when $x = \theta$ and $r = 0$.

Thus $(b, M) \in \bar{H}$ and EP is subconsistent with value M. □

If the positive cone P is closed, then this theorem has an immediate corollary obtained by reversing the roles of the primal and dual programs.

Corollary 3.4 *Let the positive cone P be closed; then* EP *is consistent with finite value M if and only if* EP* *is subconsistent with a finite subvalue M.*

We now give two examples, in both of which the primal and dual values are not the same. Theorem 3.3 shows that this can only occur when the primal value and subvalue are not equal, as is the case in our examples.

3.4.1 An example of Gale

Consider the following program:

$$\text{PG:} \quad \text{minimize} \quad x_0$$

$$\text{subject to} \quad x_0 + \sum_{i=1}^{\infty} i x_i = 1,$$

$$\sum_{i=1}^{\infty} x_i = 0,$$

$$x_i \geqslant 0, \qquad i = 0, 1, 2, \ldots .$$

We take the primal variable space to be the set of sequences with only a finite number of non-zero terms. This is called the *generalized finite sequence space* (GFSS) and we shall meet this space again in Chapter 4. Writing R^∞ for the space of all sequences, we choose to pose the problem with the following scheme of spaces:

$$\text{PG:}\begin{bmatrix}\text{GFSS}\\ R^\infty\end{bmatrix}\begin{matrix}\rightarrow\\ \leftarrow\end{matrix}\begin{bmatrix}R^2\\ R^2\end{bmatrix}\text{:PG*}$$

The only feasible solution for PG is $x_0 = 1, x_1 = x_2 = \cdots = 0$. So the value of the program is 1.

Now H is in R^3, so $\bar{H}$ can be defined from convergent sequences in H. Consider the sequence of positive $x^{(i)}$, $i = 1, 2, \ldots$, defined by

$$x_j^{(i)} = 0, \qquad i \neq j,$$

$$x_i^{(i)} = 1/i.$$

It is easy to see that the image of $x^{(i)}$ under the constraint map is $(1, 1/m)$ and the objective function takes the value 0 for each $x^{(i)}$, $i = 1, 2, \ldots$. Hence the points $(1, 1/m, 0)$ are in H, so $(1, 0, 0)$ is in $\bar{H}$. This means that the subvalue of the program is no greater than zero. In fact it is exactly zero.

To write down a dual for this program we need to find the adjoint of the constraint map; we can do this by considering the transpose of the $2 \times \infty$ matrix that represents it. We obtain the following

$$\text{PG*: maximize } y_1$$

$$\text{subject to } y_1 \leqslant 1,$$

$$ny_1 + y_2 \leqslant 0, \qquad n = 1, 2, \ldots .$$

As this last constraint holds for each value of n, we must have $y_1 \leqslant 0$. So an optimal solution to the dual problem is $y_1 = 0$, $y_2 = 0$ and the value of the program is zero. This is in accordance with the result of Theorem 3.3. The reader may like to check that if $\{x_i\}$ is the optimal solution for PG, then

$$D(\{x_i\}) = \left\{\{\xi_i\} : \sum_{i=1}^{\infty} \xi_i = 0\right\}$$

and the reduced cost functional is positive on this space, but has no positive extension to GFSS.

3.4.2 An example of Kretschmer

As a second example of this kind of behaviour consider the following inequality-constrained program:

$$\text{PK:}\quad \text{minimize} \quad \int_0^1 tx(t)\,dt + 2r$$

$$\text{subject to} \quad \int_t^1 x(s)\,ds + r \geqslant 1, \qquad t \in [0, 1],$$

$$x(t) \geqslant 0, \quad t \in [0, 1].$$

Here we take $X = Y = L_2[0, 1] \times R$ and $Z = W = L_2[0, 1]$. It is easy to see that the optimal solution to this problem is $x(t) = 0, t \in [0, 1], r = 1$, giving a value of 2. Now define functions $x^{(i)}(t)$ as follows:

$$x^{(i)}(t) = 0, \qquad t \in [0, 1 - (1/i)],$$

$$x^{(i)}(t) = i, \qquad t \in (1 - (1/i), 1],$$

and let $r^{(i)} = 0$, $i = 1, 2, \ldots$. If we apply the constraint map to $x^{(i)}$, $r^{(i)}$ (with the slacks taken as identically zero) we get a function which is 1 between 0 and $1 - (1/i)$ and then reduces linearly to 0 at 1. These functions approach the function $b(t) = 1$, $t \in [0, 1]$, in the L_2 norm. Moreover the objective function values approach 1. This is enough to show that $(b(.), 1)$ is in $\bar{H}$ and so the subvalue of the program is no more than 1.

We next formulate a dual for this problem. To do this we need to find the adjoint of the constraint map, A. Note that the image under A of the primal variable space is actually restricted to the set of absolutely continuous functions. Now

$$\begin{aligned}\langle A(x, r), w\rangle &= \int_0^1 \left(\int_t^1 x(s)\,ds\right) w(t)\,dt + r\int_0^1 w(t)\,dt \\ &= \int_0^1 \left(\int_0^t w(s)\,ds\right) x(t)\,dt + r\int_0^1 w(t)\,dt \\ &= \langle (x, r), A^*w\rangle.\end{aligned}$$

Here we have used Fubini's theorem to interchange the order of integration. Thus

$$A^*w = \left(\int_0^t w(s)\,ds, \int_0^1 w(s)\,ds\right),$$

which is in $L_2 \times R$. So the dual problem takes the following form:

$$\text{PK*:}\quad \text{maximize} \quad \int_0^1 w(t)\,dt$$

$$\text{subject to} \quad \int_0^t w(s)\,ds \leqslant t, \quad t \in [0, 1],$$

$$\int_0^1 w(s)\,ds \leqslant 2,$$

$$w(t) \geqslant 0, \qquad t \in [0, 1],$$

with w in $L_2[0, 1]$. It is not difficult to see that the optimal solution to this problem is given by $w(t) = 1, t \in [0, 1]$, giving a dual value of 1. Again this is in accordance with the result of Theorem 3.3.

3.5 CLASSIFICATION SCHEMES

In this section we discuss the possible duality states that a primal–dual pair of linear programs can take. As we shall see, information about the behaviour of the primal problem can be used to establish the behaviour of the dual problem. However, there is little or nothing in the following chapters which depends on these results, so the reader may prefer to skip this section. Recall that a program is called consistent if it has a feasible solution; we shall sometimes abreviate this to CONS. We call a problem *inconsistent* (INC) if it has no feasible solutions, *bounded* (BD) if it has a finite value and *unbounded* (UBD) otherwise. Thus a problem is unbounded if its objective function is unbounded in the direction of optimization.

We begin the classification of duality states by using the weak duality theorem (3.1). From this theorem it follows immediately that if either the primal or the dual is unbounded, then the other is inconsistent. Thus we arrive at the array of possible duality states, called a state diagram, shown in Table 1. In the table a blank square represents an impossible state.

Table 1

		Primal		
		CONS, UBD	CONS, BD	INC
	CONS UBD			1
Dual	CONS BD		2	3
	INC	4	5	6

For a finite linear program the existence of a finite optimum for the primal problem implies the existence of a finite optimum for the dual. Hence states 3 and 5 in Table 1 cannot occur. However, as later examples will show, this does not hold true for infinite linear programs.

We can take the classification of duality states one step further by introducing subconsistency (which we sometimes abbreviate to SUBC). A program which is not subconsistent we call *sub-inconsistent* (SUBINC). A program with a finite subvalue is *sub-bounded* (SUBD), a program without a finite subvalue is *sub-unbounded* (SUBUBD). Now we need a theorem on subduality which is in a sense parallel to the weak duality theorem (3.1).

Theorem 3.5 *If* EP *is subconsistent and* $(b,\alpha)\in\bar{H}$, *then for any* w *feasible for* EP*, $\alpha \geqslant \langle b, w\rangle$

Proof. Suppose the contrary. Then there is some w with $\langle b,w\rangle > \alpha$ and $c - A^*w \in P^*$. Thus for each $x \in P$, $r \geqslant 0$,

$$\langle x, c - A^*w\rangle + \langle b, w\rangle + r \geqslant \langle b, w\rangle > \alpha,$$

and so

$$\langle Ax, \ -w\rangle + \langle x, c\rangle + r \geqslant 0 > \langle b, \ -w\rangle + \alpha.$$

So (b,α) is separated from H by a hyperplane defined by $(-w, 1)$, giving a contradiction. □

We shall assume throughout the rest of this section that the positive cone P is closed, and hence that EP is its own double dual. Thus the statement of Theorem 3.5 remains true with EP and EP* interchanged and the sign of the inequality reversed. We can now add three more categories to our description of the state of a problem by splitting consistent bounded problems into those which are sub-bounded and those which are not, and by splitting inconsistent problems into three categories: (a) subconsistent and sub-bounded; (b) subconsistent and sub-unbounded; (c) sub-inconsistent.

It is easy to see from Theorem 3.5 that if a problem is sub-unbounded then its dual is inconsistent. Similarly if a problem is unbounded then its dual is sub-inconsistent. Moreover we can use Theorem 3.3 to show that a problem is subconsistent and sub-bounded if and only if its dual is consistent and bounded. These statements allow us to exclude a large number of possible duality states, and we arrive at the state diagram in Table 2, with just 9 possible duality states.

Table 2

Primal: / Dual:	CONS UBD	CONS BD SUBBD	CONS BD SUBUBD	INC SUBC SUBBD	INC SUBC SUBUBD	SUBINC
CONS UBD						1
CONS BD SUBBD		2				
CONS BD SUBUBD				3		
INC SUBC SUBBD			5			
INC SUBC SUBUBD					6	7
SUBINC	4				8	9

3.5.1 An Example of Ben-Israel, Charnes and Kortanek

The nine possible duality states of Table 2 are arrived at by a process of exclusion, but in fact all of these states actually occur in examples. We do not need to use a linear program posed in an infinite-dimensional space in order to demonstrate this. If we work in a finite-dimensional space and in addition the positive cone is finitely generated, then we get a finite linear program and only states 1, 2, 4 and 9 can occur. In this section we consider an example due to Ben-Israel, Charnes and Kortanek, posed in R^3, but with an infinitely generated positive cone, for which all the nine possible duality states can occur.

Take P to be the set of vectors making an angle of less than or equal to 45° with the vector $(1, 0, 1)$. In other words let

$$P = \{(x_1, x_2, x_3): x_1 \geqslant 0,\ x_3 \geqslant 0,\ 2x_1x_3 \geqslant x_2^2\}.$$

Then P is a closed convex cone. Moreover it is easy to see that $P^* = P$.

Consider the following linear program:

$$\begin{array}{ll} \text{minimize} & c_1x_1 + c_2x_2 \\ \text{subject to} & -x_2 + z_1 = b_1, \\ & x_3 + z_2 = b_2, \\ & x \in P, \qquad z_1 \geqslant 0, \quad z_2 \geqslant 0. \end{array}$$

This has as its dual program

$$\begin{array}{ll} \text{maximize} & b_1y_1 + b_2y_2 \\ \text{subject to} & w_1 = c_1, \\ & -y_1 + w_2 = c_2, \\ & y_2 + w_3 = 0, \\ & w \in P, \quad y_1 \leqslant 0, \quad y_2 \leqslant 0. \end{array}$$

With this primal–dual pair of problems we can obtain each of the nine possible duality states by making different choices for b and c, as follows:

State:	b_1	b_2	c_1	c_2
1	0	−1	1	0
2	0	0	1	0
3	−1	0	1	0
4	0	0	−1	0
5	0	0	0	−1
6	−1	0	0	−1
7	0	−1	0	−1
8	−1	0	−1	0
9	0	−1	−1	0

3.5.2 Homogeneous constraints

In one respect the state diagram of Table 2 is unsatisfactory: it is not one-to-one. For example, knowing that the primal is sub-inconsistent does not indicate whether the dual is consistent, subconsistent or sub-inconsistent. In this section we shall take the classification one stage further by introducing a subsidiary problem. The *modified homogeneous program* (MHEP) associated with EP is

$$\begin{aligned} \text{MHEP:}\quad \text{minimize}\quad & \langle x, \theta \rangle \\ \text{subject to}\quad & Ax = \theta, \\ & \langle x, c \rangle = -1, \\ & x \in P. \end{aligned}$$

The alternative condition for subconsistency called condition 1 in Section 3.4 is also useful here. We shall call this *asymptotic consistency* (AC). Thus EP is asymptotically consistent if b is in the closure of AP. As was pointed out in Section 3.4, if EP is subconsistent, then it is asymptotically consistent, but the reverse implication does not hold.

There is a relationship between the boundedness of a problem and whether or not its associated modified homogeneous constraints are consistent.

Theorem 3.6 (*a*) *If* EP *is consistent and bounded, the* MHEP *is inconsistent.* (*b*) *If* EP *is subconsistent, then it is sub-bounded if and only if* MHEP *is not asymptotically consistent.*

Proof. (a) If MHEP is consistent with x_0 feasible for MHEP and x_1 feasible for EP then $x_1 + \lambda x_0$ is feasible for EP provided λ is a non-negative scalar. Moreover the objective function value for EP approaches $-\infty$ as λ approaches infinity. So EP is unbounded.

(b) Suppose that EP is subconsistent. Since the set

$$H = \{(Ax, \langle x, c \rangle + r) \colon x \in P,\ r \geqslant 0\}$$

is a cone, EP is sub-bounded if and only if there is some scalar s with (b, s) not in the closure of H, that is (b, s) is not in H^{**}. As

$$H^* = \{(w, \lambda) \colon \langle Ax, w \rangle + \lambda \langle x, c \rangle + \lambda r \geqslant 0, \text{ for each } x \in P,\ r \geqslant 0\},$$

EP is sub-bounded if and only if there is some w, λ and s with

$$\langle b, w \rangle + \lambda s < 0$$

and

$$\langle Ax, w \rangle + \lambda \langle x, c \rangle + \lambda r \geqslant 0$$

for each $x \in P$ and $r \geqslant 0$. Now λ must be strictly positive, for if not then the same pair of statements is true with s replaced by any t greater than s, which contradicts the subconsistency of EP (for it implies that (b, t) is separated from H for any t).

Hence EP is sub-bounded if and only if there exists $w \in W$ and $\lambda > 0$ such that

$$\langle Ax, w\rangle + \lambda\langle x, c\rangle \geqslant 0$$

for each $x \in P$, i.e.

$$\langle Ax, w\rangle + \lambda\langle x, c\rangle \geqslant 0 > -\lambda$$

for each $x \in P$. From this it follows that $(\theta, -1)$ is separated by the hyperplane defined by (w, λ) from the set

$$\{(Ax, \langle x, c\rangle): x \in P\}$$

and MHEP is not asymptotically consistent. To make the reverse implication, observe that if MHEP is not asymptotically consistent there is some v and η with

$$\langle Ax, v\rangle + \eta\langle x, c\rangle \geqslant k > -\eta$$

for each $x \in P$. The left-hand side of this inequality is zero when $x = \theta$, and can never be negative (otherwise taking a large enough multiple of x gives a contradiction) so we can take $k = 0$ and η must be positive. □

The constraint set for MHEP is also closely related to the consistency of the dual problem EP*.

Theorem 3.7
(*a*) MHEP *is inconsistent if and only if* EP* *is asymptotically consistent.*
(*b*) MHEP *is asymptotically consistent if and only if* EP* *is inconsistent.*

Proof. (a) From the definition, we see that EP* is asymptotically consistent if c is in the closure of $P^* + A^*(W)$. This set is a convex cone, so this is equivalent to c being in $(P^* + A^*(W))^{**}$. Now

$$(P^* + A^*(W))^* = \{x \in X: \langle x, y\rangle + \langle Ax, w\rangle \geqslant 0 \text{ for all } y \in P^*, w \in W\}.$$

If x is in this set and Ax is not equal to θ, then we get a contradiction by choosing y to be θ and w so that $\langle Ax, w\rangle$ is less than zero. Hence

$$(P^* + A^*(W))^* = \{x \in P: Ax = \theta\}.$$

So EP* is asymptotically consistent if and only if $\langle x, c\rangle \geqslant 0$ for every $x \in P$ with $Ax = \theta$. It is immediate that this implies that MHEP is inconsistent. On the other hand if $\langle x, c\rangle < 0$ for some x in P with $Ax = \theta$, then by taking a positive multiple of this x we can arrange for $\langle x, c\rangle$ to be -1, and MHEP is consistent. Thus part (a) of the theorem is proved.

(b) We prove the second part of the theorem using the first part and the fact that MHEP** is MHEP (we are still assuming that P is closed). Thus MHEP is asymptotically consistent if and only if the modified homogeneous program for MHEP* is inconsistent. Now MHEP* in equality-constrained form is the program

$$\begin{aligned}&\text{maximize} && -\lambda\\ &\text{subject to} && A^*w+\lambda c+z=\theta,\\ &&& w\in W,\quad \lambda\in R,\quad z\in P^*.\end{aligned}$$

The modified homogenous program associated with MHEP* is thus

$$\begin{aligned}&\text{minimize} && \langle w,\theta\rangle+\langle z,\theta\rangle\\ &\text{subject to} && A^*w+\lambda c+z=\theta,\\ &&& \lambda=-1,\\ &&& w\in W,\quad \lambda\in R,\quad z\in P^*.\end{aligned}$$

This has a feasible set which is equivalent to that for EP* and so the second part of the theorem is established. □

We call a program which is asymptotically consistent but not subconsistent *improperly asymptotically consistent* (IAC). Before deriving the complete state diagram we need to establish one further result.

Theorem 3.8 *If a program is improperly asymptotically consistent, then its dual program is consistent.*

Proof. Suppose that EP is improperly asymptotically consistent. Then it is not subconsistent so there is certainly some s for which (b, s) can be separated from the closure of H. In other words there are w and λ with

$$\langle b,w\rangle+s\lambda<k\leqslant\langle Ax,w\rangle+\lambda\langle x,c\rangle+\lambda r$$

for all $x\in P$ and $r\geqslant 0$. By considering what happens as r becomes large we deduce that $\lambda\geqslant 0$. If λ is zero, then

$$\langle b,w\rangle<k\leqslant\langle Ax,w\rangle$$

for each $x\in P$ and hence b is not in the closure of AP. This contradicts the fact that EP is asymptotically consistent. Hence $\lambda>0$. Now

$$\langle Ax,w\rangle+\lambda\langle x,c\rangle\geqslant 0$$

for each $x\in P$; otherwise a large enough multiple of x gives a contradiction of the first inequality. Thus

$$\langle Ax,w\rangle+\lambda\langle x,c\rangle\geqslant 0>-\lambda$$

for each $x\in P$, and $(0,-1)$ is separated from the set $\{(Ax,\langle x,c\rangle): x\in P\}$, and MHEP is not asymptotically consistent. So, from Theorem 3.7(b), EP* is inconsistent. □

We now have the results we need to construct a revised state diagram in which there is a one-to-one relationship between the states of the primal and the states of the dual. This state diagram has 11 possible duality states and is given in Table 3. The state numbers correspond with those of Table 2. The nature of the modified homogenous program associated with the program EP is indicated as follows: if MHEP is consistent, we call the program HC; if MHEP is asymptotically consistent but not consistent we call the program HA; and if MHEP is not asymptotically consistent we call it HI.

Table 3

PRIMAL: / DUAL:				CONS				INC						
				UBD		BD		SUBC			SUBINC			
						SUB BD	SUB UBD	SUB BD	SUBUBD		IAC			
				HA	HC	HI	HA	HI	HA	HC		HI	HA	HC
CONS	UBD		HA								1A			
CONS	UBD		HC									1B		
CONS	BD	SUBBD	HI			2								
CONS	BD	SUBUBD	HA					3						
INC	SUBC	SUBBD	HI				5							
INC	SUBC	SUBUBD	HA						6					
INC	SUBC	SUBUBD	HC										7	
INC	IAC			4A										
INC	SUBINC		HI		4B									
INC	SUBINC		HA							8				
INC	SUBINC		HC											9

3.6 CONDITIONS FOR THE ABSENCE OF A DUALITY GAP

A crucial property of finite linear programs is that the program and its dual have the same value. As we have seen, this is not always the case for infinite linear programs, and we will consider in some detail conditions under which this property holds. We will use these conditions a number of times in the following chapters, and the reader should look at Chapters 4, 6 and 7 for examples of the application of the theorems given in this section. In particular Section 4.2 gives applications of Theorems 3.10, 3.11 and 3.19, Section 4.4 gives applications of

Theorem 3.13 and Corollary 3.18, while both Section 6.5 and Section 7.2 have applications of Theorem 3.15.

It must be admitted, however, that the various conditions do not have as wide a range of applicability as one would wish. We shall illustrate our discussion with a particular example:

$$\text{P0:}\quad \text{minimize} \quad \int_0^1 c(t)x(t)\,\mathrm{d}t$$

$$\text{subject to} \quad \int_t^1 a(s)x(s)\,\mathrm{d}s \geqslant b(t),$$

$$x(t) \geqslant 0, \qquad t \in [0, 1].$$

A dual problem is

$$\text{P0*:}\quad \text{maximize} \quad \int_0^1 b(t)w(t)\,\mathrm{d}t$$

$$\text{subject to} \quad \int_0^t a(s)w(s)\,\mathrm{d}s \leqslant c(t),$$

$$w(t) \geqslant 0, \qquad t \in [0, 1],$$

where we take all the spaces involved as $L_2[0, 1]$.

This pair of programs may or may not have the same value. For example consider the problem P0 with $a(t) = c(t) = 1, t \in [0, 1]$. If $b(t) = 1$ throughout the interval, then P0 has no feasible solution while P0* has a value of 1. On the other hand if $b(t) = 1$, $t \in [0, \frac{1}{2}]$, and $b(t) = 0$ otherwise, then both P0 and P0* have a value of 1. Ideally we should like to know conditions on the functions a, b and c which ensure that the two programs have the same value.

If a program and its dual have the same (finite) value, then we say that there is *no duality gap* for the program. A related condition, called *strong duality*, is that the primal and dual programs have the same value and in addition this value is achieved in each program. In other words there exist x and y, feasible for the primal and dual respectively, with $\langle x, c \rangle = \langle b, y \rangle$. From Theorem 3.3 we see that there is no duality gap for a program when its value and subvalue are equal (and finite).

3.6.1 Closedness conditions

The most obvious conditions that ensure that the value and subvalue of a program are equal are those involving the closedness of some set. The following theorem is an easy consequence of the definitions.

Theorem 3.9 *If* EP *is consistent, with finite value, and H is closed, then there is no duality gap for* EP.

Proof. From the definition of subvalue, and the closedness of H (see Section 3.5), the subvalue of EP is equal to its value. Hence from Theorem 3.3, EP* is consistent and its value is equal to the value of EP. □

There is also a counterpart to this theorem which uses the set D rather than H.

Theorem 3.10 *If* EP *is consistent, with finite value, and D is closed, then there is no duality gap for* EP.

Proof. We only need to show that the subvalue of EP is equal to its value. Suppose that the value of EP is m and there is some $\alpha < m$ with (b, α) in $\bar{H}$. Then from the definition of m, (b, α) is not in D and can be separated from it by a hyperplane, defined by (w, λ) say. Thus

$$\langle b, w \rangle + \lambda\alpha < \langle Ax, w \rangle + \lambda \langle x, c \rangle$$

for all $x \in P$. Now

$$m = \inf\{\mu\colon (b, \mu) \in D\}.$$

As D is closed, $(b, m) \in D$ and so $\lambda > 0$. Thus

$$\langle b, w \rangle + \lambda\alpha < \langle Ax, w \rangle + \lambda \langle x, c \rangle + \lambda r$$

for all $x \in P$ and $r \geqslant 0$. Hence (b, α) is separated from H by this hyperplane and cannot lie in $\bar{H}$. This is a contradiction and shows that the subvalue of EP is m. □

The fact that closedness of either D or H can be used to establish the absence of a duality gap should not be surprising. It is only the lower boundaries of these sets which are important to the theorem, and these are essentially the same. However, neither condition is strictly stronger than the other. Examples can be found where D is closed but H is not, and vice versa.

It is clear that there are versions of Theorems 3.9 and 3.10 for the inequality-constrained problem. The sets D and H have to be replaced by sets D' and H', defined as follows

$$D' = \{(Ax - z, \langle x, c \rangle)\colon x \in P,\ z \in Q\},$$

$$H' = \{(Ax - z, \langle x, c \rangle + r)\colon x \in P,\ z \in Q,\ r \geqslant 0\}.$$

When the positive cone is closed there are similar results for the dual problems, EP* and IP*.

If we wish to apply these results to problem P0, then it is necessary to consider whether or not the set

$$D' = \{(f, \lambda)\colon f(t) \leqslant \int_t^1 a(s)x(s)\,ds,\ \lambda = \int_0^1 c(t)x(t)\,dt, \quad \text{for some } x(t) \geqslant 0\},$$

is closed (or the similar set H'). Such conditions will be extremely hard to verify directly.

3.6.2 Interior point conditions

There are a number of ways of guaranteeing the closedness of the sets D or H, but the most usual conditions for this involve the existence of interior points in some set. In this section several such interior point conditions are discussed, but we shall derive them directly as conditions ensuring the absence of a duality gap, rather than as conditions ensuring that either D or H is closed.

The conditions of the next theorem sound complicated, but they turn out to be surprisingly easy to apply.

Theorem 3.11 *Suppose that there is some $\tau(Z, W)$-neighbourhood B of θ with $B+b$ contained in AP. If* EP *has a finite value and there is some γ such that for each $z \in b+B$ there is an $x \in P$ with $Ax = z$ and $\langle x, c \rangle \leqslant \gamma$, then there is no duality gap for* EP.

Proof. As usual we wish to show that the value and subvalue of EP are the same. Suppose that EP has a value of α and a subvalue of β with $\beta = \alpha - \varepsilon$, for some $\varepsilon > 0$. Recall that, as H is convex, the closure of H is the same under the weak topology and under the Mackey topology. As $(b, \beta) \in \overline{H}$ and $(1/n)B$ is a neighbourhood of θ for each n, we can find points $(z_n, \lambda_n) \in H$ with $z_n \in b + (1/n)B$ and $\lambda_n \leqslant \alpha - \varepsilon/2$. So there are points $x_n \in P$ with $Ax_n \in b + (1/n)B$ and

$$\langle x_n, c \rangle \leqslant \alpha - \varepsilon/2.$$

We may choose B as one of the base of neighbourhoods, and hence balanced. Thus $n(b - Ax_n)$ is in B. So, by assumption, there is a sequence $\{u_n : n = 1, 2, \ldots, \}$, $u_n \in P$, such that

$$Au_n = b - n(Ax_n - b)$$

and with $\langle u_n, c \rangle \leqslant \gamma$. Let

$$z = (nx_n + u_n)/(n+1).$$

Then z is a convex combination of two points in P and $Az = b$, so z is feasible and $\langle z, c \rangle \geqslant \alpha$. But

$$\langle u_n, c \rangle = (n+1)\langle z, c \rangle - n\langle x_n, c \rangle.$$

Hence

$$\langle u_n, c \rangle \geqslant (n+1)\alpha - n(\alpha - \varepsilon/2) = \alpha + n\varepsilon/2,$$

which for n large enough contradicts the choice of u_n. □

If, as in Chapter 2, we define the optimal value functional by

$$\omega(z) = \inf\{\langle x, c^* \rangle : Ax = z,\ x \geqslant 0\}, \qquad z \in Z,$$

then the conditions of this theorem are equivalent to the $\tau(Z, W)$-continuity of ω. For, in terms of the value functional, the conditions state that there is a γ with $\omega(z) \leqslant \gamma$ on $b + B$. The convexity of ω then ensures that it is $\tau(Z, W)$-continuous at b. A continuity condition of this sort is usually referred to as a *stability* condition.

Note that interior point conditions involving the Mackey topology are the weakest that can be given in topologies consistent with a particular dual pair. Thus for example in the above theorem we only need to find some topology on Z consistent with the dual pair (Z, W) and such that there is a neighbourhood of the origin with the required properties.

Theorem 3.12 *Let X and Z be Banach spaces, Y and W their dual spaces and let A map X onto Z. If* EP *has a finite value and there is an x_0 in the interior of P with $Ax_0 = b$, then there is no duality gap for* EP.

Proof. We can choose E to be an open neighbourhood of x_0 which lies within P such that $\langle x, c \rangle$ is bounded on E. Then b is in AE which is within AP. Now AE is an open neighbourhood of b by the open mapping theorem (see for example Lang (1983), p. 208), so we can apply Theorem 3.11 to obtain the result. □

There are also interior point conditions for the inequality-constrained problem and its dual. In fact for these problems the conditions take on a particularly simple form.

Theorem 3.13 *If* IP *has a finite value and there is an x_0 in P with $Ax_0 - b$ in the $\tau(Z,W)$-interior of Q, then there is no duality gap for* IP.

Proof. Let $Ax_0 - b = z_0$ where N is a $\tau(Z, W)$-neighbourhood of θ with $z_0 + N$ in Q. Write A' for the map from $X \times Z$ to Z defined by $A'(x, z) = Ax + z$. Then $A'(x_0, z_0) = b$ and $A'(x_0, z_0 + N) = b + N$. So b is in the interior of $A'(P, Q)$. If the program IP is put in the form EP by replacing A by A', X by $X \times Z$ and P by $P \times Q$, then we can apply Theorem 3.11 to give the result. □

Corollary 3.14 *Suppose P and Q are closed. If* IP* *has a finite value and there is a w_0 in Q^* with $c - A^* w_0$ in the $\tau(Y,X)$-interior of P^*, then there is no duality gap for* IP.

Now we turn again to the example P0. This is an inequality-constrained problem, so we look first at Theorem 3.13. Unfortunately this theorem cannot apply to our problem in its original form

$$\text{P0: } \begin{bmatrix} L_2[0, 1] \\ L_2[0, 1] \end{bmatrix} \begin{matrix} \rightarrow \\ \leftarrow \end{matrix} \begin{bmatrix} L_2[0, 1] \\ L_2[0, 1] \end{bmatrix} \text{: P0*}$$

as Q, the positive cone in L_2, has no interior points (even in the Mackey topology).

However, the choice of L_2 as the space in which to pose the problem was fairly arbitrary. This leads us to consider a formulation in which the primal constraint space (which contains Q) is L_∞; this is the only one of the L_p-spaces to have a positive cone with interior points (in its norm topology). We might consider the following formulation:

$$\text{P0:}\begin{bmatrix} L_1[0,1] \\ L_\infty[0,1] \end{bmatrix}\begin{matrix}\rightarrow \\ \leftarrow\end{matrix}\begin{bmatrix} L_\infty[0,1] \\ L_1[0,1] \end{bmatrix}:\text{P0*}$$

In fact this will not work, as L_1 is not the dual of L_∞ (though the reverse is true) and the Mackey topology $\tau(L_\infty, L_1)$ gives a positive cone in L_∞ with no interior.

The difficulty of getting a positive cone with interior points is partly explained by a result we shall give later (Theorem 3.16). The same problems occur in trying to apply Theorem 3.12. Here we require an interior point of the positive cone in the primal variable space. The only candidate Banach spaces are thus $L_\infty[0,1]$ and $C[0,1]$ (the continuous functions on $[0,1]$). Because the inequality-constrained problem must be put into equality-constrained form, and the constraint map must be surjective, we end up with the same space for the primal constraints. Hence the dual variable space W must be the dual Banach space for either L_∞ or C, and thus will not be a function space.

3.6.3 Boundedness conditions

One way of showing that one of the sets H or D is closed is to impose conditions which bound the feasible set in terms of the right-hand side of the constraints, b. In fact we do not need to bound the whole feasible set, but only a subset of it containing the optimal solution. This is the approach that is taken in the next theorem.

Theorem 3.15 *Let X, Y, W, and Z be normed spaces with X the dual of Y and W the dual of Z. For any given b, let $F(b)$ be the set of feasible solutions. Suppose that the positive cone P is closed. If* EP *has a finite value and there exist sets $B(b)$, with $B(b)$ contained in $F(b)$ for each b, and such that*

(a) there is some M_1 and M_2 independent of b, with $\|u\| \leqslant M_1 + M_2\|b\|$ for each $u \in B(b)$,
(b) for each $x \in F(b)$ there is an $x' \in B(b)$ with $\langle c, x'\rangle \leqslant \langle c, x\rangle$,
then there is no duality gap for EP.

Proof. We need to establish only that H is closed. As H is convex it is sufficient to show that H is closed in the norm topology. Suppose that (b_n, r_n) is a sequence in H converging in the norm to (b, r). Thus there is $x_n \in F(b_n)$, $n = 1, 2, \ldots,$ with $\langle c, x_n\rangle \leqslant r_n$. By the conditions of the theorem we may choose $x_n \in B(b_n)$. Then

$$\|x_n\| \leqslant M_1 + M_2\|b_n\|.$$

Since b_n approaches b we can choose M_3 with $\|b_n\| \leqslant M_3$ for all n, and so $\|x_n\|$ is bounded. As the unit ball in X is $\sigma(X, Y)$-compact (this is Alaoglu's theorem; see for example Robertson and Robertson (1973), p. 62), we can choose a subnet of x_n which converges in the weak topology, to x, say. As P is closed and A is continuous, $x \in P$ and $Ax = b$, so $x \in F(b)$. Also $\langle c, x \rangle \leqslant r$, so $(b, r) \in H$. □

This is a result which can be applied to problems of the same general type as P0, as we demonstrate in Chapter 7. However, it turns out that it is not applicable to P0 itself. It looks as though it will be easiest to apply the theorem to P0*, and it will be convenient to take the simplest case, when $a(t) = 1, 0 \leqslant t \leqslant 1$. We then wish to find M_1 and M_2 with

$$\|(w, z)\| \leqslant M_1 + M_2\|c\|,$$

whenever $z \geqslant 0$, $w \geqslant 0$ and

$$\int_0^t w(s)\,ds + z(t) = c(t), \qquad t \in [0, 1]. \tag{7}$$

At first sight this seems to be easy, but the spaces X and W must be duals of Y and Z respectively. This implies either our original L_2 formulation or (in equality constrained form)

$$\text{P0:}\begin{bmatrix} L^2_\infty[0, 1] \\ L^2_1[0, 1] \end{bmatrix} \begin{matrix} \rightarrow \\ \leftarrow \end{matrix} \begin{bmatrix} L_1[0, 1] \\ L_\infty[0, 1] \end{bmatrix}\text{: P0*}$$

However the constraint (7) does not bound the supremum of w, only its integral. So we cannot use this formulation, nor indeed any L_p space with $p > 1$, for the primal variables.

3.6.4 Compactness conditions

The material of this section requires a little more mathematical background, and is somewhat peripheral to the main development. We need the following result.

Proposition 7 *If (X, Y) is a dual pair of vector spaces, then the Mackey topology on X has a base of neighbourhoods at the origin consisting of the sets A^0 where A is any balanced convex $\sigma(Y, X)$-compact set in Y.*

A cone P in X is said to have a *compact base* if there is some $y \in Y$ and $\alpha > 0$ such that the set

$$S(P, y, \alpha) = \{x \in P: \langle x, y \rangle = \alpha\}$$

is $\sigma(X, Y)$-compact and spans P. It is easy to show that this set spans P if and only if $\langle x, y \rangle > 0$ for all $x \in P$, $x \neq \theta$. We call such a y *strictly positive* in Y. A cone P in X is said to have a *compact sole* if there is a compact set U in P such that θ is not in U and U spans P. Thus a cone with a compact base certainly has a compact sole, though the reverse implication does not hold.

Theorem 3.16 *If (X,Y) is a dual pair and P is a $\sigma(X,Y)$-closed cone in X, then the following conditions are equivalent:*

(1) *P has a compact base,*
(2) *P^* has a non-empty interior in the Mackey topology $\tau(Y,X)$.*

Proof. (a) Take y_0 in the interior of P^* in the $\tau(Y,X)$ topology, so there is some $\tau(Y,X)$-neighbourhood of the origin V with y_0+V in P^*. Hence there is some $\sigma(X,Y)$-compact, convex, balanced set A in X and $\delta>0$ with $y_0+\delta A^0$ in P^*. Consider x in $S(P,y_0,\delta)$; as $x\in P$,

$$\langle x,y_0\rangle+\langle x,y\rangle\geqslant 0$$

for each $y\in V$, and so $\langle x,y\rangle\geqslant-\delta$. But V is balanced so x is in δV^0. Thus x is in $\delta(\delta A^0)^0$, which is the set A. Hence $S(P,y_0,\delta)$ is a closed subset of a compact set and is compact. Now suppose that $\langle x,y_0\rangle=0$ for some $x\in P$. Then $\langle x,y\rangle\geqslant 0$ for each $y\in V$ and hence $x=\theta$. Thus y_0 is strictly positive in Y and $S(P,y_0,\delta)$ spans P.

(b) Suppose that $S(P,y_0,\varepsilon)$ is compact and spans P. Let K be the convex hull of $S(P,y_0,\varepsilon)$ and $-S(P,y_0,\varepsilon)$. This is balanced and $\sigma(X,Y)$-compact (being the convex hull of two compact sets). Hence $V=\delta K^0$ is a $\tau(Y,X)$ neighbourhood of θ. If $x\in S(P,y_0,\varepsilon)$ and $y\in V$, then $-\varepsilon\leqslant\langle x,y\rangle\leqslant\varepsilon$ and so

$$\langle x,y_0+y\rangle=\varepsilon+\langle x,y\rangle\geqslant 0.$$

But $S(P,y_0,\varepsilon)$ spans P, so y_0+V is in P^* and hence y_0 is in the Mackey interior of P^*. □

When P has a compact sole there is a simple set of conditions which ensure that the set D is closed; and hence that there is no duality gap for EP when it has a finite value.

Theorem 3.17 *If P is a cone with compact sole and there is no $x\in P$, other than θ, with $Ax=\theta$ and $\langle x,c\rangle=0$, then the set D is closed.*

Proof. As P has a compact sole we can choose a compact set U not containing θ and such that

$$P=\{\lambda x: x\in U,\lambda\geqslant 0\}.$$

Now suppose that $\{x_\alpha\}$ is a net in P such that $\{(Ax_\alpha,\langle x_\alpha,c\rangle)\}$ converges to (z,β). Let $x_\alpha=\lambda_\alpha u_\alpha$, where $\lambda_\alpha\geqslant 0$ and $u_\alpha\in U$. As U is compact we can choose a convergent subnet of $\{u_\alpha\}$, converging to u say. Now we consider two cases depending on the behaviour of the corresponding subnet of $\{\lambda_\alpha\}$.

(a) If λ_α is bounded for $\alpha\geqslant\alpha_0$ for some α_0, then we can choose a subnet from $\{\lambda_\alpha\}$ converging to λ say. Then

$$Ax_\alpha=\lambda_\alpha Au_\alpha$$

and so $z = \lambda u$. But

$$\langle x_\alpha, c \rangle = \lambda_\alpha \langle u_\alpha, c \rangle$$

so $\beta = \lambda \langle u, c \rangle$. Now $\lambda u \in P$ and hence $(z, \beta) \in D$.

(b) If λ_α is not bounded for α greater than any α_0, then we can choose a subnet with λ_α approaching ∞. Thus $Au_\alpha - z/\lambda_\alpha$ converges to Au, but $\lambda_\alpha Au_\alpha - z$ converges to θ, so $Au = \theta$. Similarly $\langle u_\alpha, c \rangle - \beta/\lambda_\alpha$ converges to $\langle u, c \rangle$, but $\lambda_\alpha \langle u_\alpha, c \rangle - \beta$ converges to 0, so $\langle u, c \rangle = 0$. As u is in U, $u \neq \theta$ which gives a contradiction. □

We remark that the condition that P has a compact sole is quite restrictive, but the other conditions of the theorem will usually hold in practice. The following corollary is an immediate consequence of Theorems 3.16 and 3.17.

Corollary 3.18 *If P is a closed cone, P^* has a non-empty interior in $\tau(Y, X)$ and there is no $x \in P$ other than θ with $Ax = \theta$ and $\langle x, c \rangle = 0$, then the set D is closed.*

The corresponding theorem to 3.17 for the inequality-constrained problem is given below. Notice that this theorem does not require any condition on Q, as it would if it had been deduced directly from Theorem 3.17. However, its proof is straightforward and is left as an exercise for the reader.

Theorem 3.19 *If P is a cone with compact sole and there is no $x \in P$, other than θ, with Ax in Q and $\langle x, c \rangle = 0$, then the set D' is closed.*

As we mentioned above, the condition that P has a compact sole is restrictive, and will not be satisfied by the problem P0 whatever the choice of function space. Thus this particular problem has resisted the whole battery of techniques available to us. It is however instructive to attempt to apply these results, and doing so at least demonstrates the care that is needed in the choice of function spaces in which to pose this type of problem.

3.7 THE EXISTENCE OF OPTIMAL SOLUTIONS

We have not so far addressed the question of whether there is a feasible solution which achieves the optimum value. As is conventional, the programs involved have been written as 'maximize' or 'minimize', but in many cases there may be no feasible solution which achieves the optimum value. If there is such a feasible solution it is called an *optimal solution.* A program with an optimal solution is called *solvable.* In this section we shall investigate conditions under which a program is guaranteed to be solvable.

For convenience we shall only consider the program EP. First consider the case that the feasible region is $\sigma(X, Y)$-compact. Then, since a continuous real-valued function is bounded and attains its bounds on a compact set, the program EP is solvable.

Theorem 3.20 *If P is closed, X and Y are normed spaces such that X is the dual of Y and the feasible region is bounded in the norm of X, then* EP *is solvable.*

Proof. Recall the theorem of Alaoglu (see for example Robertson and Robertson (1973)) that the unit ball in X is compact in the $\sigma(X,Y)$ topology. As P is closed and A is continuous, the feasible region is $\sigma(X,Y)$-closed and lies within a compact set. Hence the feasible region is compact and the result follows. □

The following result gives a more complicated way in which compactness can be used to ensure solvability. First define for each real number r, the set $K(r)$ of feasible points with an objective function value of r, i.e.

$$K(r) = \{x \in P : Ax = b,\ \langle x, c\rangle = r\}.$$

Theorem 3.21 *Suppose that P is closed and that* EP *has a finite value. If there exist real numbers $r_1 > r_2$ such that $K(r_1)$ is non-empty and $K(r_2)$ non-empty and compact, then* EP *is solvable.*

Proof. Let v be the value of EP. If $v = r_2$, then any element of $K(r_2)$ is an optimal solution and we are done. So suppose that $v < r_2$. Let x_1 be an element of $K(r_1)$. Choose r strictly between v and r_2. From the definition of v, $K(r)$ is non-empty. If $x \in K(r)$, then an appropriate convex combination of x and x_1 is in $K(r_2)$. Hence x is an element of the set $\lambda_1 x_1 + \lambda_2 K(r_2)$ for

$$\lambda_1 = (r - r_2)/(r_1 - r_2).$$

$$\lambda_2 = (r_1 - r)/(r_1 - r_2).$$

Define the set $K[v, r]$ as follows:

$$K[v, r] = \{x \in P : Ax = b,\ v \leqslant \langle x, c\rangle \leqslant r\},$$

for $v < r < r_2$. As P is closed and A is continuous, $K[v, r]$ is closed. Moreover $K[v, r]$ is contained in the set

$$\{(1-\lambda)x_1 + \lambda K(r_2) : \lambda \in [1, \alpha]\},$$

where

$$\alpha = (r_1 - v)/(r_1 - r_2).$$

So $K[v, r]$ is contained in a compact set (the continuous image of $[1, \alpha] \times K(r_2)$). Thus it is compact. But

$$K(v) = \bigcap_{v < r < r_2} K[v, r],$$

and so is non-empty by the finite intersection property. This establishes the theorem. □

These compactness conditions are not the only way to ensure that the program EP is solvable.

Theorem 3.22 *If* EP *has a finite value and either of the sets D or H are closed then* EP *is solvable.*

Proof. If D is closed, then $D \cap L$ is closed when considered as a subset of R. Hence, if v is the value of EP, the point (b, v) is in $D \cap L$. Thus there is some $x \in P$ with $Ax = b$ and $\langle x, c \rangle = v$. The proof is similar if H is closed. □

It is interesting that the conditions of Theorem 3.22 also ensure strong duality for EP. Note that we can use the conditions of some of the results of Section 3.6 to ensure the closedness of one of these sets (e.g. 3.15, 3.17, 3.18, 3.19).

Having discussed the existence of optimal solutions it is appropriate to ask whether or not there are optimal solutions which are basic. It is not hard to find linear programs with optimal solutions but no extreme point (i.e. basic) optimal solutions. Consider the following example.

$$
\begin{aligned}
&\text{minimize} && x_1 \\
&\text{subject to} && \sum_{i=1}^{\infty} (x_i)/2^i \leqslant 1, \\
& && \sum_{i=j}^{\infty} x_i \geqslant 1, \qquad j = 1,2,\ldots, \qquad\qquad (8) \\
& && x_i \geqslant 0, \qquad i = 1,2,\ldots .
\end{aligned}
$$

The solution $x_1 = 0, x_i = 1, i = 2,3, \ldots$, is feasible and clearly optimal. But it is not basic; indeed the feasible region has no extreme points. The constraints (8) imply that any feasible solution has an infinite number of non-zero components. Now if x_j and $x_k, j \neq k$, are non-zero, then changing x_j by $\pm 2^j \delta$ and x_k by $\mp 2^k \delta$ produces two other feasible solutions for δ chosen small enough.

We can achieve the same type of behaviour with a finite-dimensional problem if we allow either the positive cone to be not closed, or the feasible region to be unbounded. Indeed the example we have given has, in some sense, an unbounded feasible region; though each component of x is individually bounded.

Fortunately, there is a well-known theorem (see e.g. Holmes (1975) p. 74) which states that if A is a compact set in a space X and f is any lower semi-continuous concave function on X, then f attains its minimum on A at an extreme point of A. Applied to our problem, this result ensures that a linear program has a basic optimal solution whenever the feasible region is $\sigma(X, Y)$-compact.

3.8 NOTES

The study of duality theory for infinite dimensional linear programs started with a paper of Duffin (1956). He posed the problem in a slightly different way to that we have adopted. He restricted X to be the topological dual of Y and W to be the topological dual of Z. Duffin proves a version of Theorem 3.3 as his main result. Kretschmer (1961), five years later, adopted the paired space formulation which we have used. He re-proved Duffin's results in this setting. Kretschmer showed that the set H for the problem IP* is closed if the interior point condition of Theorem 3.13 holds. This establishes that there is no duality gap for IP under these conditions, and also that the dual problem is solvable, hence giving a slight strengthening of our result. He also gave several examples including that of Section 3.5.2. Following on from Kretschmer's work are the papers by Yamasaki (1968) and Nakamura and Yamasaki (1978), who also treat the problem in a paired space setting. These authors give a great variety of conditions for strong duality. Further work on duality theory, especially as it applies to continuous linear programs, can be found in Pomerol (1985).

Another strand in the theory of infinite dimensional linear programs goes back to a lengthy paper of Hurwicz (1958). He gives a generalization of a lemma of Farkas (1902). Farkas' lemma is essentially the following result: the system

$$Ax = b, \qquad x \geqslant 0, \quad x \in R^n,$$

has a solution if and only if

$$A^{\mathrm{T}} y \geqslant 0, \quad y \in R^m \qquad \text{implies that} \qquad b^{\mathrm{T}} y \geqslant 0,$$

where A is an $m \times n$ matrix, and $b \in R^m$. This lemma can be treated as a starting point for the theory of linear optimization, and from it duality theorems for linear programming can be derived. Versions of Farkas' lemma for infinite-dimensional spaces are also given by Braunschweiger and Clark (1962), Ben-Israel and Charnes (1968), Ben-Israel *et al.* (1969), Craven (1972), Levine and Pomerol (1974), Kortanek (1977), Zalinescu (1978), Schirotzek (1981) and Komiya (1981). Ben-Israel *et al* were the first to give a classification of the relationships which can exist between primal and dual problems. This classification was extended in the survey article by Kallina and Williams (1971), and also by Kortanek (1974).

Borwein (1983b) derives a wide range of linear-programming duality results by treating our LP formulation as a special case of the generalized linear program, which has the constraints $Ax = b, x \geqslant \theta$, replaced by $b \in H(x)$ where $H: X \to Y$ is a convex process (a set-valued function whose graph is a convex cone). A similar formulation is given by Nozawa (1981).

The linear programming formulation we have given can be generalized by allowing P and Q to be arbitrary convex sets. Such problems are treated by Fan (1969) and (for finite dimensions) by Ben Israel *et al.* (1971). It is also possible to consider extensions which replace the objective function $\langle c, x \rangle$ with a linear map f from X to an ordered vector space V. Then, rather than minimize $\langle c, x \rangle$, we minimize $f(x)$ with respect to the ordering on V. This is the direction taken in the papers by Ritter (1969a, 1969b) and Zowe (1975).

Theorem 3.13 is well known; a proof can be found in Aubin (1971), and a similar result is proved by Fan (1965), who uses it to derive a set of interior point conditions for strong duality. A similar but weaker set of conditions is derived by Allen (1977).

The example of Section 3.5.1 is due to Gale and appears in a paper by Van Slyke and Wets (1968). This paper, though dealing with a much wider class of problem than the linear programs we consider, contains some relevant material. The same can be said of the book by Ponstein (1980).

Problems

3.1 Check the formulation of EP** given in Section 3.3.1.
3.2 Find the sets D, H and L for the example of Section 3.2.1.
3.3 When put into standard equality constrained form, the dual program for the example of Kretschmer is as follows:

$$\begin{aligned} &\text{minimize} \quad -\int_0^1 w(t)\,\mathrm{d}t \\ &\text{subject to} \quad \int_0^t w(s)\,\mathrm{d}s + u(t) = t, \qquad t \in [0, 1], \\ &\qquad\qquad\quad \int_0^1 w(s)\,\mathrm{d}s + v = 2, \\ &\qquad\qquad\quad w(t), \quad u(t), \quad v \geqslant 0, \qquad t \in [0, 1]. \end{aligned}$$

Show that, if the primal variable space is taken to be the set of $w(t)$ in $L_\infty[0, 1]$, $u(t)$ differentiable with $u(0) = 0$ and v in R, then $w(t) = 1$, $u(t) = 0$, $v = 1$ is a non-degenerate basic feasible solution. Show that the reduced-cost functional for this solution is positive.
3.4 Justify the statement made in Section 3.6.2 that only the 11 states shown in the state diagram of Table 3 are possible.
3.5 Modify the example of Section 3.5.1 to obtain the two possible duality states (out of 11) which are not illustrated there.
3.6 Call a cone C in a space X *finitely generated* if there are $x_1, x_2, \ldots, x_n$ in X with

$$C = \{(\Sigma \lambda_i x_i) : \lambda_i \geqslant 0\}.$$

Assuming that P is finitely generated and that all finitely generated cones in the primal constraint space Z are $\sigma(Z, W)$-closed, show that there is no duality gap for EP if it has a finite value. (This result applies to the usual finite linear program.)
3.7 Use a finite-dimensional linear program to show that the conditions of Theorem 3.12 are not necessary for strong duality.
3.8 Prove Theorem 3.19.
3.9 Consider the problem P0 with $a(t) = c(t) = 1$, $t \in [0, 1]$ and $b(t) = 1 - t$, $t \in [0, 1]$. Show that P0 and P0* have the same value, P0 is solvable and P0* is not solvable. Explain why Theorem 3.20 does not apply to P0*.
3.10 Consider the L_1 approximation problem, introduced in Problem 2.9, posed in the form IP with the following scheme of spaces

$$\begin{bmatrix} C(0, 1) \times R^n \\ M_r(0, 1) \times R^n \end{bmatrix} \begin{matrix} \rightarrow \\ \leftarrow \end{matrix} \begin{bmatrix} (C(0, 1))^2 \\ (M_r(0, 1))^2 \end{bmatrix}$$

where $M_r(0, 1)$ is the space of regular Borel measures on (0, 1). Show that the appropriate dual problem is

$$\text{maximize} \quad \int_0^1 f(s)\,\mathrm{d}(\mu_1 - \mu_2)$$

$$\text{subject to} \quad \int_0^1 g_i(s)\,\mathrm{d}(\mu_1 - \mu_2) = 0, \qquad i = 1,2, \ldots, n,$$

$$\mu_1 + \mu_2 = \mu,$$

$$\mu_1 \geqslant 0, \qquad \mu_2 \geqslant 0,$$

where μ is Lebesgue measure on the line. Show that the dual problem is solvable and that there is no duality gap.

4

Semi-infinite Linear Programs

4.1 INTRODUCTION

In the last two chapters we dealt with the theory of infinite-dimensional linear programming in general. We come now to apply this theory to specific classes of problem. In the hope that there will be fewer complications when either the number of constraints or the number of variables remains finite, than for a fully infinite program, it is natural to start with semi-infinite linear programs. In fact, as we shall see even semi-infinite linear programs possess a very rich structure, both in their duality theory and in simplex extensions.

There are two major classes of semi-infinite linear programs; those where the infinite index set is countable and those where the index set is a subset of Euclidean space (such as a line segment or rectangle). We shall call them *countable* and *continuous* semi-infinite programs respectively. In this chapter we give a fairly comprehensive account of both types of problem. A number of useful conditions which ensure that the primal and dual programs have the same value can be obtained as simple applications of the results of Chapter 3. In addition, we describe simplex-like algorithms for both the countable and continuous semi-infinite programs. In fact in each case we discuss two different algorithms: one for the primal problem and one for the dual problem.

One of the most important applications of semi-infinite programming is to approximation theory. As an illustration of some of the ideas of this chapter we discuss the uniform approximation problem, and show how the well known Remez exchange algorithm is related to the simplex extensions we consider.

4.2 THE COUNTABLE SEMI-INFINITE PROGRAM AND ITS DUAL

A number of formulations are possible for the semi-infinite linear programming problem. The most usual is as follows:

$$\begin{aligned} \text{SIP:}\quad & \text{minimize} && c^{\mathrm{T}} x \\ & \text{subject to} && a(s)^{\mathrm{T}} x \geqslant b(s),\ s \in S, \\ & && x \in R^{n}. \end{aligned}$$

Here c and $a(s)$ are n-vectors and S is some index set. If S is a finite set then the semi-infinite program reduces to the finite linear program. In this section we take S to be countably infinite. Thus the semi-infinite program takes the form

$$\begin{aligned} \text{SIP1:} \quad & \text{minimize} && c^{\mathrm{T}} x \\ & \text{subject to} && a_i^{\mathrm{T}} x \geqslant b_i, \qquad i = 1,2,\ldots, \\ & && x \in R^n. \end{aligned}$$

Problem SIPI is an inequality-constrained program. In the notation of Chapter 3, it is in the form IP with the primal variable space X and its positive cone P both equal to R^n. The primal constraint space must be some sequence space with the usual positive cone. Before discussing the dual problem we need to specify more precisely the primal constraint space and the dual variable space. As usual we have several choices open to us. The main restriction in practice is that the sequences $\{a_i^{\mathrm{T}} x\}$ and $\{b_i\}$ must lie in the primal constraint space.

We shall choose Z, the primal constraint space, to be the space of all sequences, R^∞, and W, the dual variable space, to be the space of all sequences with only a finite number of non-zero terms. This latter space is usually called the generalized finite sequence space, or GFSS. A bilinear form is defined on Z and W in the natural way by letting $\langle z, w \rangle$ be $\Sigma\, z_i w_i$. We can summarize this choice of spaces as follows:

$$\text{SIP1:} \begin{bmatrix} R^n \\ R^n \end{bmatrix} \begin{matrix} \rightarrow \\ \leftarrow \end{matrix} \begin{bmatrix} R^\infty \\ \text{GFSS} \end{bmatrix} : \text{SIP1*}$$

It is not difficult to write down the adjoint of the constraint map. As $P^* = \{\phi\}$ and $Q^* = \{w : w_i \geqslant 0,\ i = 1,2,\ldots\}$, we arrive at the following dual problem.

$$\begin{aligned} \text{SIP1*:} \quad & \text{maximize} && \Sigma\, w_i b_i \\ & \text{subject to} && \Sigma\, w_i a_i = c, \qquad (1) \\ & && w_i \geqslant 0, \text{ and only a finite number of the } w_i \text{ non-zero.} \end{aligned}$$

In many ways this dual problem has a more convenient structure than the primal. It is of the equality-constrained form EP with variable space GFSS, constraint space R^n and the usual positive cone in GFSS. The dual of SIP1* is easily seen to be SIP1. This statement also follows from observing that the positive cone in R^∞ is closed in its usual topology. In fact the algebraic dual of GFSS is R^∞, so SIP1 is actually the algebraic dual of SIP1*, though the reverse is not the case.

It is easy to check that $\sigma(Z, W)$ is the usual product topology on R^∞. Perhaps because GFSS is the dual of R^∞ with its usual topology, the choices we have made for Z and W are the standard ones made in the literature. In fact these choices of spaces allow us to derive duality results which encompass almost all of those obtainable even with different choices of spaces. It seems to be hard, if not impossible, to find examples of countable semi-infinite programs which have a duality gap in this formulation, but have no duality gap when we take W to be a wider class of sequences. Note that if W is extended, then Z may have to be restricted if there is to be a bilinear form defined on $Z \times W$. However, a restriction of Z has no effect on the value of the primal problem SIP1.

Having formulated the primal and dual problems, we seek conditions which will ensure that their values coincide. The three theorems given below give three sets of conditions, all of which are quite easy to derive from the more general theorems given in Chapter 3. First we need to define two sets:

$$M_A = \{\Sigma w_i a_i : w_i \geqslant 0 \text{ and all but a finite number zero}\},$$

$$M_B = \{(\Sigma w_i a_i, \Sigma w_i b_i) : w_i \geqslant 0 \text{ and all but a finite number zero}\}.$$

Note that M_A and M_B are cones in R^n and R^{n+1} respectively; they are often called *moment cones*. If we write A for the constraint map of SIP1*, then

$$M_A = \{Ax : x \geqslant \theta\},$$

$$M_B = \{(Ax, \langle x, b \rangle) : x \geqslant \theta\}.$$

Thus M_B is just the set we called D in Section 3.5, and we derive the following theorem immediately from Theorems 3.10 and 3.22.

Theorem 4.1 *If the value of* SIP1* *is finite and* M_B *is closed, then the values of* SIP1 *and* SIP1* *coincide and* SIP1* *is solvable.*

The set M_A is the image of the positive cone under the constraint map for SIP1*. So SIP1* is consistent exactly when c is in M_A. The following result gives an interior-point condition for the absence of a duality gap.

Theorem 4.2 *If* SIP1* *has a finite value and* c *is in the interior of* M_A *then the values of* SIP1 *and* SIP1* *coincide.*

Proof. We can choose $n+1$ points in M_A so that c is in the interior of their convex hull. Thus

$$c = \sum_{i=1}^{n} \lambda_i y_i,$$

where $y_i \in M_A$, $\Sigma \lambda_i = 1$ and $1 > \lambda_i > 0$. Now $y_i = A(w^{(i)})$, $i = 0, 1, \ldots, n$, where the $w^{(i)}$ are some positive elements of GFSS. Thus c is the image under A of $\Sigma \lambda_i w^{(i)}$ and a neighbourhood of c is obtained as the image under A of the set

$$\{\Sigma \mu_i w^{(i)} : \mu_i \geqslant 0, \Sigma \mu_i = 1\},$$

which is contained in P. Moreover $\langle b, \cdot \rangle$ is clearly bounded on this set. These are the properties required in order to apply Theorem 3.11 to SIP1*, which gives the result. □

Finally we have a result derived from Theorems 3.19 and 3.22.

Theorem 4.3 *If* SIP1 *has a finite value and there is no* x, *other than zero, with* $c^{T}x = 0$ *and* $a_i^{T}x \geqslant 0$, $i = 1,2, \ldots$, *then the values of* SIP1 *and* SIP1* *coincide and* SIP1 *is solvable.*

Proof. For SIP1 the positive cone P is just R^n, which has a compact sole (e.g. the surface of the unit sphere). Hence we can apply Theorem 3.19 to show that the set D' is closed. Here D' is a set in $R^{\infty} \times R$ defined by

$$D' = \{(a_1^{T}x - z_1, a_2^{T}x - z_2, \ldots; c^{T}x): x \geqslant 0, z_i \geqslant 0, i = 1,2, \ldots\}.$$

Thus the result follows from Theorems 3.10 and 3.22. □

Another way to approach the question of existence of duality gaps, which is particularly appropriate for the countable semi-infinite problem, involves looking at a truncated version of SIP1 where we consider only the first k constraints. This problem has the form

$$\begin{aligned} \text{SIP}_k\text{:} \quad & \text{minimize} & & c^{T}x \\ & \text{subject to} & & a_i^{T}x \geqslant b_i, \qquad i = 1,2, \ldots, k, \end{aligned}$$

which has as its dual

$$\begin{aligned} \text{SIP}_k{}^*\text{:} \quad & \text{maximize} & & \sum_{i=1}^{k} b_i w_i \\ & \text{subject to} & & \sum_{i=1}^{k} w_i a_i = c, \\ & & & w_i \geqslant 0 \end{aligned}$$

As these are finite programs the values of SIP_k and its dual coincide (at least when one of them is given as finite). Let their common value be v_k.

Now any feasible solution for $\text{SIP}_k{}^*$ is also a feasible solution for SIP1* (by setting $w_{k+1} = w_{k+2} = \cdots = 0$). So for each k, v_k is a lower bound on $V(\text{SIP1}^*)$, the value of the program SIP1*. We might guess that v_k approaches $V(\text{SIP1}^*)$ as k gets large and, as we now show, this is indeed the case. There is some sequence of points $w^{(j)}, j = 1,2, \ldots$ in GFSS such that $\langle b, w^{(j)} \rangle$ approaches $V(\text{SIP1}^*)$ from below. Let $N(j)$ be chosen so that $w_i^{(j)} = 0$ for $i \geqslant N(j)$. Then the first $N(j)$ components of $w^{(j)}$ are feasible for $\text{SIP}_{N(j)}{}^*$ and so $v_{N(j)}$ is greater than $\langle b, w^{(j)} \rangle$. Since the sequence $\{v_k\}$ is bounded above by $V(\text{SIP1}^*)$, it must converge to $V(\text{SIP1}^*)$. The import of all this is that one may view a duality gap between SIP1 and SIP1* as a failure of the sequence $\{V(\text{SIP}_k)\}$ to converge to $V(\text{SIP1})$.

It is important to realize that all conditions given in this section can be checked within the context of a finite-dimensional space. We shall illustrate this remark with an example. Notice that we have already met one example of a semi-infinite linear program with a duality gap, in Section 3.5.1. In fact the example there is quite similar to the one we give below.

4.2.1 An example of Karney

Consider the following program

$$\begin{aligned} \text{P1:} \quad & \text{minimize} \quad && x_1 \\ & \text{subject to} && x_1 \geqslant -1, \\ &&& -x_2 \geqslant 0, \\ &&& x_1 - (1/i)x_2 \geqslant 0, \qquad i = 3,4,\ldots. \end{aligned}$$

This has optimal solution $x_1 = x_2 = 0$ with a value 0. The dual problem is

$$\begin{aligned} \text{P1*:} \quad & \text{maximize} \quad && -w_1 \\ & \text{subject to} && w_1 + w_3 + w_4 + \cdots = 1, \\ &&& -w_2 - (w_3/3) - (w_4/4) - \cdots = 0, \\ &&& w_i \geqslant 0 \text{ and only a finite number non-zero.} \end{aligned}$$

This has optimal solution $w_1 = 1, w_i = 0, i = 2,3,\ldots$, giving a value -1. Hence there is a duality gap.

For this problem the moment cone M_A is generated by

$$\{(1,0), (0,-1), (1,-1/3), (1,-1/4), \ldots\}.$$

The point (1,0) (which is c) is contained in this cone but is on its boundary. The moment cone M_B is generated by

$$\{(1,0,-1), (0,-1,0), (1,-1/3,0), (1,-1/4,0), \ldots\}$$

and is not closed; it has an open boundary along the x_1 axis. Finally observe that the direction $(0,-1)$ is perpendicular to c and each a_i has a non-negative component in this direction. Thus we have confirmed that none of the three theorems above will apply in this case.

This also provides a counterexample to a conjecture that we might make; namely that the existence of an x_0 in R^n with $a_i^T x_0 > b_i$, $i = 1, 2, \cdots$ would ensure that there was no duality gap. It is tempting to conclude that this remark follows from Theorem 3.13. However the positive cone in R^∞ does not have any interior points (in the appropriate Mackey topology), and so we cannot apply this theorem.

Finally we remark that if we regard the dual problem for this example as a primal problem, we have an example of a program with a duality gap between it and its algebraic dual. This demonstrates that it is not always possible to close a duality gap by including in the space of dual variables more linear forms on the primal constraint space.

4.3 SIMPLEX EXTENSIONS FOR COUNTABLE SEMI-INFINITE PROGRAMS

In this section we discuss simplex-like algorithms for the solution of the countable semi-infinite program and its dual. In doing so we shall make considerable use of the general concepts introduced in Chapter 2. Since the dual problem is quite unlike the primal problem, it is not surprising that the simplex extension for the dual problem should be rather different to that for the primal problem. Usually a strong duality result holds. This means that we are able to choose which of the two problems, primal or dual, we wish to solve. Thus we shall obtain two different and alternative algorithmic approaches to these problems, and we discuss these in turn.

4.3.1 An algorithm for the dual problem

We start by identifying the basic solutions for the dual countable semi-infinite problem. Let $w = \{w_i\}$ be feasible for SIP1*. For any $u \in$ GFSS we shall write $\text{supp}(u)$ for the support of u, i.e. the finite set $\{i : u_i \neq 0\}$. Then

$$B(w) = \{u \in \text{GFSS}: \text{supp}(u) \subset \text{supp}(w)\}.$$

For w to be a basic solution, and hence an extreme point of the feasible set, we require $B(w) \cap N(A) = \{\theta\}$, where A is the constraint map for SIP1*. This happens if and only if $\Sigma\, u_i a_i = 0$ and $\text{supp}(u) \subset \text{supp}(w)$ together imply $u_i = 0$ for each i. In other words a feasible w is a basic solution if and only if the set $\{a_i : i \in \text{supp}(w)\}$ is linearly independent. An immediate implication of this is that there are no more than n elements in $\text{supp}(w)$.

In order for a simplex-like algorithm to be effective we need a simple representation of the reduced-cost functional. This makes it important that this functional be defined on the whole of the primal variable space, in other words that the current basic feasible solution be non-degenerate. For this problem a feasible solution w is non-degenerate when $\{a_i : i \in \text{supp}(w)\}$ spans R^n and hence there are exactly n elements in $\text{supp}(w)$. Given a basic feasible solution w, we write $\hat{A}$ for the matrix whose columns are the a_i, $i \in \text{supp}(w)$, $\hat{w}$ and $\hat{b}$ for the vectors with elements $w_i, b_i, i \in \text{supp}(w)$. Then if w is non-degenerate, $\hat{A}$ is square and invertible with $\hat{A}\hat{w} = c$. Hence $\hat{w} = \hat{A}^{-1}c$.

The next step is to identify the reduced-cost functional for a basic non-degenerate solution w. This is an element of R^∞ which we call w^*. Recall from Chapter 2 that the reduced-cost functional x^* for a basic feasible solution x is given by

$$\langle y, x^* \rangle = \langle y, c^* \rangle - \langle A_B{}^{-1} A y, c \rangle.$$

Thus in this case, for $y \in$ GFSS,

$$\begin{aligned} \langle y, w^* \rangle &= \Sigma\, y_i b_i - \hat{b}^{\mathrm{T}} \hat{A}^{-1} (\Sigma\, y_i a_i) \\ &= \Sigma\, y_i (b_i - a_i{}^{\mathrm{T}} (\hat{A}^{-1})^{\mathrm{T}} \hat{b}). \end{aligned}$$

So w^* can be identified with the element of R^∞ whose ith component is

$$b_i - a_i^{\mathrm{T}}(\hat{A}^{-1})^{\mathrm{T}}\hat{b}.$$

Note that w_i^* is zero for $i \in \text{supp}(w)$. As SIP1* is posed as a maximization problem, we know from Chapter 2 that a basic feasible non-degenerate w is optimal if and only if w^* is non-positive (i.e. each component of $w^* \leqslant 0$). This happens when $(\hat{A}^{-1})^{\mathrm{T}}\, \hat{b}$ is feasible for SIP1.

We are now ready to describe the pivot step in this setting. This pivot step moves along an extreme ray of the set of feasible perturbations; it is an example of the type of pivot step described in a general setting in Theorem 2.7.

Suppose then that w^* is not non-positive, in other words there is some k with

$$b_k > a_k(\hat{A}^{-1})^{\mathrm{T}}\,\hat{b}.$$

Consider setting $w_k = \delta > 0$. As w is basic there is only one way to retain feasibility (i.e. to satisfy (1)) with supp(w) unaltered except for the addition of the index k. In fact, writing primes for the new w, we have

$$\sum_{\text{supp}(w)} w_i' a_i + \delta a_k = c.$$

So, with a slight abuse of notation, we may write

$$w' = \hat{w}' + \delta e_k,$$

where

$$\hat{w}' = \hat{A}^{-1}(c - \delta a_k), \tag{2}$$

and we write e_k for the vector with kth element 1 and other elements zero. Thus

$$\begin{aligned} \Sigma\, b_i w_i' - \Sigma\, b_i w_i &= \hat{b}\hat{w}' - \hat{b}\hat{w} + \delta b_k \\ &= \delta b_k - \delta \hat{b}^{\mathrm{T}} \hat{A}^{-1} a_k \\ &> 0. \end{aligned}$$

In other words w' is an improvement on w. Moreover w' is feasible for δ small enough.

To specify a simplex-like algorithm we need to make w' a basic solution, so that the process of generating improved solutions can be continued. We can do this by increasing δ until some w_i' hits zero. In fact we need to choose δ as

$$\min \{w_i / [\,(\hat{A}^{-1})^{\mathrm{T}} a_k]_i\},$$

where the minimization is taken over $i \in \text{supp(w)}$ such that the denominator is positive. Because supp(w) is finite, the minimum will be achieved.

Another perspective on this algorithm is obtained if it is viewed as operating on the primal problem, SIP1. There it moves through a succession of infeasible solutions to the primal (as the dual simplex method does in finite LP). Viewed in this way the algorithm takes the following form: choose an initial set of n inequality constraints in SIP1; solve the primal problem with these constraints alone, and if there are no other constraints broken then this solution is optimal for

SIP1, otherwise choose a constraint which is broken and add it to the constraint set. Then re-solve the problem; one of the constraints will now become inactive and it can be dropped from the constraint set before restarting the process.

4.3.2 An algorithm for the primal problem

Having discussed a simplex-like procedure for the dual problem, we now look for such a procedure for the primal problem. We must start by considering the basic solutions. Suppose x is feasible for SIP1. Define $\text{constr}(x)$ to be the set of active constraints. In other words

$$\text{constr}(x) = \{i : a_i^{\mathrm{T}} x = b_i\}.$$

This set will turn out to play a role that is dual to that of $\text{supp}(w)$ in the previous section. Define the matrix $\tilde{A}$ and the vector $\tilde{b}$ in the same way as $\hat{A}$ and $\hat{b}$ but with $\text{supp}(w)$ replaced by $\text{constr}(x)$.

We add slacks to SIP1 to turn it into an equality constrained program, obtaining

$$\begin{aligned} &\text{minimize} && c^{\mathrm{T}} x \\ &\text{subject to} && a_i^{\mathrm{T}} x - z_i = b_i, \qquad i = 1, 2, \ldots, \\ & && z_i \geqslant 0. \end{aligned} \tag{3}$$

A solution to this problem is a pair (x, z). Naïvely, we might expect

$$B(x, z) = \{(u, v) : u \in R^n, v \in R^\infty, v_j = 0, j \in \text{constr}(x)\},$$

as there is no sign constraint on x. If the sequence $\{z_j\}$ is bounded away from zero for j large enough, this will be correct. Some thought will show, however, that if $\liminf z_j = 0$, $B(x, z)$ will be a smaller space than this. Suppose for the moment that $\liminf z_j > 0$. Then x is basic if $B(x, z)$ intersects the null space of the constraint map only at the origin. This is true if and only if the rows of $\tilde{A}$ are linearly independent. Of course, (x, z) will still be basic if this condition holds, even if $\liminf z_j = 0$. However, (x, z) will then be degenerate. If the rows of $\tilde{A}$ are linearly independent, $\text{constr}(x)$ must have at least n elements. If, in addition, $\liminf z_j > 0$, then (x, z) will be non-degenerate if $\text{constr}(x)$ has exactly n elements. In this case $\tilde{A}$ is square invertible. We have $\tilde{A}^{\mathrm{T}} x = \tilde{b}$, so

$$x = (\tilde{A}^{-1})^{\mathrm{T}} \tilde{b} \tag{4}$$

if x is a non-degenerate basic solution.

Suppose that x is a non-degenerate basic feasible solution with associated slacks $\{z_i\}$. The reduced cost x^* is an element of $R^n \times R^\infty$. The result of its operation on an element of the same space, say (u, v), is given by $c^{\mathrm{T}} u - c^{\mathrm{T}} (\tilde{A}^{-1})^{\mathrm{T}} d$, where d is the vector of elements of $\{a_i^{\mathrm{T}} u - v_i\}$ indexed by $\text{constr}(x)$. Thus

$$\langle (u, v), x^* \rangle = c^{\mathrm{T}} (\tilde{A}^{-1})^{\mathrm{T}} \tilde{v}$$

where $\tilde{v}$ is composed of the elements of v indexed by $\text{constr}(x)$. So a non-degenerate basic feasible solution x is optimal if and only if $\tilde{A}^{-1}c$ is non-negative.

As for the dual problem, the pivot step moves along an extreme ray of the set of feasible perturbations, and is exemplified in the result of Theorem 2.7. In detail, we suppose that the kth element of $\tilde{A}^{-1}c$ is less than zero. Consider setting the slack variable indexed by the kth element of constr(x) to the value $\delta > 0$. Then, from (4), x takes on the value

$$(\tilde{A}^{-1})^{\mathrm{T}}(\tilde{b}+\delta e_k),$$

and hence, from (3), z_i changes by $\delta a_i^{\mathrm{T}}(\tilde{A}^{-1})^{\mathrm{T}}e_k$. The change in the value of the objective function is given by

$$\delta c^{\mathrm{T}}(\tilde{A}^{-1})^{\mathrm{T}}e_k$$

and this is positive. Moreover the new x is feasible for δ chosen small enough.

To obtain a basic solution we take δ just large enough for some other slack to become zero. If this is not possible then the problem does not have a finite optimum. In fact we must take

$$\delta = \inf\{-z_i/[\tilde{A}^{-1}a_i]_k\},$$

where the minimization is taken over i for which the denominator is negative. If this infimum is not achieved, then the new set of slacks will have $\lim\inf(z_i) = 0$.

It is clear that the implementation of either the dual or primal simplex extension outlined above would be very difficult. In fact any algorithm for SIP1 will be difficult to implement; it is hard even to check feasibility when there are a countably infinite number of constraints. Paradoxically the same problems do not necessarily occur when the constraint set is uncountably infinite. This is the case for the continuous semi-infinite program discussed in the next sections.

4.4 THE CONTINUOUS SEMI-INFINITE PROGRAM AND ITS DUAL

In this section we consider the semi-infinite program when the index set S is taken as a subset of R^k. Such problems are called continuous semi-infinite programs. We shall take $k = 1$ and $S = [0, 1]$. This is partly for convenience of exposition; much of what follows remains true when S is any compact subset of Euclidean space.

Our problem now takes the form

$$\begin{array}{lll} \text{SIP2:} & \text{minimize} & c^{\mathrm{T}}x \\ & \text{subject to} & a(s)^{\mathrm{T}}x \geqslant b(s), \qquad s \in [0, 1]. \end{array}$$

Throughout this section and the next we shall assume that the functions a and b are continuous, which is a natural assumption in this context. The continuous semi-infinite program is an optimization problem posed over R^n with a linear objective function and a convex feasible region. In fact most problems with these characteristics can be represented as continuous semi-infinite programs.

Write $C[0, 1]$ for the space of continuous functions on $[0, 1]$. As before we need to make a choice of spaces before we can write down a dual problem for

SIP2. In the notation of Chapter 3 this problem is an inequality-constrained program with primal variable space (and positive cone) R^n, and primal constraint space $C[0, 1]$ with the usual positive cone. The natural choice for the dual variable space is $M_r[0, 1]$, the space of regular Borel measures on $[0, 1]$, as this is the dual space of $C[0, 1]$ with its usual topology. We can express all this in the following diagram:

$$\text{SIP2:} \begin{bmatrix} R^n \\ R^n \end{bmatrix} \begin{matrix} \rightarrow \\ \leftarrow \end{matrix} \begin{bmatrix} C[0, 1] \\ M_r[0, 1] \end{bmatrix} : \text{SIP2*}$$

Now

$$\begin{aligned} \langle x, A^*, w \rangle &= \langle Ax, w \rangle \\ &= \int_0^1 a(s)^{\mathrm{T}} x \, \mathrm{d}w(s) \\ &= x^{\mathrm{T}}\left(\int_0^1 a(s) \, \mathrm{d}w(s)\right). \end{aligned}$$

So

$$A^* w = \int_0^1 a(s) \mathrm{d}w(s).$$

Also $P^* = \{\theta\}$, and Q^* is the set of positive measures in $M_r[0, 1]$. Hence the dual problem for SIP2 is as follows:

$$\begin{aligned} \text{SIP2*:} \quad & \text{maximize} && \int_0^1 b(s) \, \mathrm{d}w(s) \\ & \text{subject to} && \int_0^1 a(s) \, \mathrm{d}w(s) = c, \\ & && w \geqslant 0, \; w \in M_r[0, 1]. \end{aligned}$$

As with the countable semi-infinite program, the positive cone in the primal space, $C[0, 1]$, is closed in the appropriate topology, and so SIP2 is its own double dual.

We wish to find conditions under which there will be no duality gap for this pair of problems. Again we start by defining moment cones in R^n and R^{n+1} as follows:

$$M_A = \left\{ \int_0^1 a(s) \, \mathrm{d}w(s) : w \geqslant 0, \; w \in M_r[0, 1] \right\}$$

$$M_B = \left\{ \int_0^1 \bar{a}(s) \, \mathrm{d}w(s) : w \geqslant 0, \; w \in M_r[0, 1] \right\},$$

where we write $\bar{a}(s)$ for $(a(s), b(s))$. The three theorems given in Section 4.2 all have counterparts in this situation, and the theorem below gathers together these results.

Theorem 4.4 *Any of the three conditions below are sufficient to ensure that the values of* SIP2 *and* SIP2* *are the same:*

(*a*) SIP2* *has a finite value and* M_B *is closed,*
(*b*) SIP2* *has a finite value and c is in the interior of* M_A.
(*c*) SIP2 *has a finite value and there is no x, other than* 0, *with* $c^T x = 0$ *and* $a(s)^T x \geqslant 0$, $s \in [0, 1]$.

Proof. Inspection of the proofs of Theorems 4.1, 4.2 and 4.3 show that they apply unchanged in this case. □

There are also two conditions for the absence of a duality gap which have no counterpart in the countable case. First note that the positive cone in $C[0, 1]$ has a non-empty interior (of functions which are strictly positive throughout $[0, 1]$). So we can apply Theorem 3.13 to SIP2 and obtain the following result.

Theorem 4.5 *If there is an* x_0 *such that* $a(s)^T x_0 > b(s)$ *for each* $s \in [0,1]$ *and* SIP2 *has finite value, then the values of* SIP2 *and* SIP2* *are the same.*

In addition to this result there is a weaker condition which ensures the closure of the moment cone M_B, and thus allows the application of Theorem 4.4(a). We begin by establishing a characterisation of the moment cone M_B. Let

$$G = \{\bar{a}(s) : s \in [0, 1]\}.$$

Theorem 4.6 M_B *is the convex cone generated by G.*

Proof. G is a compact set, so the convex hull of G is also compact (Schaefer (1971), p. 66). Suppose $g = \int_0^1 \bar{a}(s)\,dw(s)$ is in M_B, with $\int_0^1 dw(s) = K$. We shall show that $(1/K)g$ is in the convex hull of G.

We have

$$(1/K)g = \int_0^1 \hat{a}(s)\,dw(s)/K$$

$$= \lim_{n \to \infty} g_n$$

where

$$g_n = \sum_{i=1}^{n} \left(\bar{a}(i/n) \int_{(i-1)/n}^{i/n} (1/K)\,dw(s) \right).$$

This follows because the integral can be evaluated as the limit of a Riemann sum when the integrand is continuous. Now g_n is a convex combination of points in G. Because the convex hull of G is compact, $(1/K)g$ is also in the convex hull of G, which establishes the result. □

The convex cone generated by a compact set which misses the origin is closed (Holmes (1975), p. 76). Thus we obtain the following corollary, involving a condition which is very easy to check.

Corollary 4.7 *If* SIP2* *has a finite value and there is no* s_0 *in* $[0,1]$ *with*

$a(s_0) = 0$ and $b(s_0) = 0$, then the values of SIP2 *and* SIP2* *are the same and* SIP2* *is solvable.*

Proof. This result follows immediately from Theorems 4.6, 4.4 and 3.18. □

Theorem 4.6 can also be used to characterize a subset of the feasible solutions to SIP2* within which there must be an optimal solution. In fact the following result shows that we can without loss of generality restrict the solutions of SIP2* to atomic measures with finite support. This in turn shows that if there is an optimal solution, then there is a basic optimal solution (from Theorem 2.5).

Theorem 4.8 *If there is a feasible solution for* SIP2* *with value c_0, then there is a feasible solution with mass concentrated on a finite number of points which achieves the same value.*

Proof. From the conditions of the theorem we know that (c,c_0) is in M_B. Thus, using Theorem 4.6,

$$(c,c_0) = \sum_{i=1}^{k} \lambda_i \bar{a}(s_i),$$

where $\lambda_i \geqslant 0$, $s_i \in [0,1]$, $i = 1, 2, \ldots, k$ and, from Caratheodory's theorem, k can be taken as $n+2$. Hence we may take w as a measure with masses λ_i concentrated at points s_i, and w is then a feasible solution with value c_0, as required. □

Despite the variety of duality results we have given, it is not hard to find examples of continuous semi-infinite programs with duality gaps. We finish this section by giving one such example.

Consider the problem

$$\begin{aligned} &\text{minimize} && x_1 \\ &\text{subject to} && s x_1 + s^2(1-s)x_2 \geqslant -s^2, \qquad s \in [0, 1]. \end{aligned}$$

For s near enough to zero, any negative value of x_1 makes the constraint impossible to satisfy. Hence the solution $x_1 = x_2 = 0$ is optimal, giving a value of 0.

The dual problem is as follows

$$\begin{aligned} &\text{maximize} && \int_0^1 (-s^2)\,\mathrm{d}w(s) \\ &\text{subject to} && \int_0^1 s\,\mathrm{d}w(s) = 1, \\ &&& \int_0^1 s^2(1-s)\,\mathrm{d}w(s) = 0, \\ &&& w \geqslant 0, \qquad w \in M_r[0, 1]. \end{aligned}$$

As w is positive, the second constraint forces every feasible solution to the dual to have its mass concentrated at 0 and 1. The first constraint ensures that there is a mass of 1 at the point 1. Hence the dual problem has a value of -1, achieved by every feasible solution. It is instructive to check the conditions of the various theorems of this section, and observe how they fail to be satisfied for this problem.

4.5 SIMPLEX EXTENSIONS FOR CONTINUOUS SEMI-INFINITE PROGRAMS

In this section we describe simplex-like algorithms for the solution of the continuous semi-infinite program and its dual. These are similar to the two simplex extensions presented in Section 4.3. However, there are some crucial differences; it is interesting that these appear only with the primal problem. The algorithm for the dual problem goes through almost unchanged from Section 4.3.1 and we shall discuss this first.

4.5.1 An algorithm for the dual problem

As we saw in Theorem 4.8 of the previous section, there is no loss of generality in taking a solution of SIP2* as atomic with masses at a finite number of points, and we now make this assumption. For any atomic measure u in $M_r[0,1]$, we write $\text{supp}(u)$ for the finite set $\{s: u(s) \neq 0\}$, of points where u has some mass. Then if an atomic measure w is feasible for SIP2*,

$$B(w) = \{u \in M_r[0, 1]: \text{supp}(u) \subset \text{supp}(w)\}.$$

The condition for w to be basic is that this set intersect the null space of the constraint map, $N(A)$, only at the origin. Here

$$N(A) = \left\{u \in M_r[0, 1]: \int_0^1 a(s)\, du(s) = 0\right\}.$$

So w is basic if and only if the vectors $a(s_i)$, $s_i \in \text{supp}(w)$ are linearly independent. So, if w is basic, $\text{supp}(w)$ can contain no more than n elements. A basic feasible w is non-degenerate if the image under the constraint map of $B(w)$ is the same as the image of the whole of $M_r[0, 1]$. This will be the case if the vectors $a(s_i)$, $s_i \in \text{supp}(w)$, span R^n, or in other words if $\text{supp}(w)$ contains exactly n elements.

Suppose now that w is a basic non-degenerate feasible solution for SIP2*. As in Section 4.3.1, we write $\hat{A}$ for the matrix with columns $a(s_i)$, $s_i \in \text{supp}(w)$; $\hat{w}$ and $\hat{b}$ for the vectors with elements $w(s_i)$ and $b(s_i)$, $s_i \in \text{supp}(w)$.

Then $\hat{A}\hat{w} = c$ and, as w is non-degenerate, $\hat{A}$ is square invertible, so

$$\hat{w} = \hat{A}^{-1}c.$$

By an exact analogy with Section 4.3.1, the reduced-cost functional w^* is given by

$$w^*(s) = b(s) - a(s)^{\mathrm{T}}(\hat{A}^{-1})^{\mathrm{T}}\hat{b},$$

and is in $C[0, 1]$. Note that $w^*(s_i) = 0$, $s_i \in \operatorname{supp}(w)$. We have, from the general result of Chapter 2, that w is optimal if and only if w^* is non-positive, i.e. if and only if $(\hat{A}^{-1})^{\mathrm{T}}\hat{b}$ is feasible for SIP2.

Now suppose that w is not optimal, then there is some s_0 with

$$b(s_0) > a(s_0)^{\mathrm{T}}(\hat{A}^{-1})^{\mathrm{T}}\hat{b}.$$

Consider a new feasible solution for SIP2*, w' say, with

$$\operatorname{supp}(w') = \operatorname{supp}(w) \cup \{s_0\},$$

and $w'(s_0) = \delta$. Then

$$\hat{A}\hat{w}' + \delta a(s_0) = c,$$

where $\hat{w}'$ is that part of w' within $\operatorname{supp}(w)$. So

$$\hat{w}' = \hat{A}^{-1}c - \delta\hat{A}^{-1}a(s_0).$$

Thus

$$\int_0^1 b(s)\,\mathrm{d}w'(s) - \int_0^1 b(s)\,\mathrm{d}w(s) = \delta b(s_0) - \delta\hat{b}^{\mathrm{T}}\hat{A}^{-1}a(s_0)$$

$$> 0.$$

Hence w' is an improvement on w. Moreover w' is feasible for SIP2* if δ is chosen small enough.

Again, as in Section 4.3.1, we wish to make w' a basic solution. To do this we must take δ just large enough for one of the elements of $\hat{w}'$ to become zero. In other words

$$\delta = \min\{\hat{w}_i/[\hat{A}^{-1}a(s_0)]_i\}, \tag{5}$$

where the minimization is taken over $i \in \operatorname{supp}(w)$ such that the denominator is positive.

In practice the right choice of s_0 is important to the effectiveness of this algorithm. In fact s_0 should be chosen to maximize, or nearly maximize, the value of $w^*(s_0)$.

4.5.2 An algorithm for the primal problem

In this section we shall assume that the functions a and b are analytic, and so have Taylor-series expansions at each point. With the addition of slacks the problem SIP2 can be rewritten

$$\begin{array}{lll} \text{minimize} & c^{\mathrm{T}}x & \\ \text{subject to} & a(s)^{\mathrm{T}}x - z(s) = b(s), & s \in [0, 1], \\ & x \in R^n, \quad z(s) \geqslant 0, & s \in [0, 1]. \end{array}$$

Note that z is also an analytic function.

Suppose that x is feasible for SIP2 and z is the associated slack variable function. Define constr(x), a subset of $[0, 1]$, by

$$\text{constr}(x) = \{s: a(s)^{\mathrm{T}}x = b(s)\}.$$

This is the set of points where z is zero. As z is analytic, either it has a finite number of zeros in $[0, 1]$, or it is identically zero. We shall assume that the function $b(s)$ is not in the space spanned by $a_i(s)$, $i = 1, 2, \ldots, n$, and so z cannot be identically zero. Thus we may take constr(x) to be a set with a finite number of elements, say $s_1, s_2, \ldots, s_m$.

Before investigating the basic solutions for this problem, we give a fundamental lemma characterizing $B(z)$. Note that $z(s_i) = 0$, $i = 1,2, \ldots, m$. Moreover, as z is non-negative, z must just touch the constantly zero line at these points. Hence $z'(s_i) = 0$ for those s_i not equal to 0 or 1. This motivates the following definition. Let $d(i)$, $i = 1,2, \ldots, m$ be defined so that

$$z^{(k)}(s_i) = 0, \qquad k = 0,1, \ldots, d(i),$$
$$z^{(k)}(s_i) \neq 0, \qquad k = d(i)+1, \quad i = 1,2, \ldots, m.$$

When s_i is neither 0 or 1, it is not hard to see from the Taylor-series expansion of z that $d(i)$ must be an odd number (we have already seen that $d(i) \geqslant 1$).

Lemma 4.9 *Let $v \in C[0, 1]$ be analytic, then $v \in B(z)$ if and only if $v^{(j)}(s_i) = 0$, $j = 0,1, \ldots, d(i)$, $i = 1,2, \ldots, m$.*

Proof. Recall that $v \in B(z)$ if there is some λ with $z(s) \pm \lambda v(s) \geqslant 0$, $s \in [0,1]$. Write $g_\lambda(s)$ for $z(s) + \lambda v(s)$. Consider a fixed i and let $k = d(i)+1$. Then

$$g_\lambda(s_i+\delta) = \lambda v(s_i) + \delta\lambda v^{(1)}(s_i) + \frac{\delta^2}{2}\lambda v^{(2)}(s_i) + \cdots$$
$$+\frac{\delta^k}{k!}z^{(k)}(s_i) + \frac{\delta^{k+1}}{(k+1)!}z^{(k+1)}(s_i) + \cdots.$$

We require both $g_\lambda(s_i+\delta)$ and $g_{-\lambda}(s_i+\delta)$ to be non-negative. Hence the terms in δ^j must be zero for $j = 0,1, \ldots, d(i)$. This proves the result in one direction.

Now suppose that $v^{(j)}(s_i) = 0$, $j = 0,1, \ldots, d(i)$, $i = 1,2, \ldots, m$. Consider a fixed i and write k for $d(i)+1$. Then

$$g_\lambda(s_i+\delta) = \frac{\delta^k}{k!}(z^{(k)}(s_i) + \lambda v^{(k)}(s_i)) + O(\delta^{k+1}). \tag{6}$$

Consider first the case when s_i is neither 0 nor 1. Note that k is an even number and that $z^{(k)}(s_i) > 0$; otherwise z itself would not be non-negative. Hence there is some $\eta > 0$ such that, for λ small enough in modulus, $g_\lambda(s_i+\delta) \geqslant 0$, $|\delta| < \eta$. In the case that $s_i = 0$ or 1, then δ in (6) is restricted in sign. However, essentially the same argument will apply. Thus we can find similar open intervals around each of the s_i in which g_λ is non-negative for λ small enough in modulus.

Let S be the remainder of the interval $[0, 1]$ when these open intervals have been removed. We can bound v in modulus on S (it is a continuous function on a compact set). Similarly we can bound z away from zero on S. Hence, for λ small enough in modulus, $g_\lambda(s) > 0$ for $s \in S$. This establishes the other half of the result. □

Using this lemma we can characterize the basic solutions for SIP2. For this problem we have

$$B(x, z) = \{(u, v) : u \in R^n, v \in C[0, 1] \text{ and } v \in B(z)\},$$
$$N(A) = \{(u, v) : u \in R^n, v(s) = a(s)^{\mathrm{T}} u, s \in [0, 1]\}.$$

Now, x is a basic solution if $B(x, z) \cap N(A) = \{0\}$. Thus x is basic if $a(s)^{\mathrm{T}} u$ in $B(z)$ implies $u = 0$.

Write $\tilde{A}$ for the matrix whose columns are $a^{(j)}(s_i)$, $j = 0, 1, \ldots, d(i)$, $i = 1, 2, \ldots, m$. Then we can use Lemma 4.9 to obtain the following result.

Theorem 4.10 *If x is feasible for* SIP2, *then x is basic if and only if the rows of $\tilde{A}$ are linearly independent.*

For any infinitely differentiable function $v(s)$, $s \in [0, 1]$ we shall write $\tilde{v}$ for the vector formed from the elements $v^{(j)}(s_i)$, $j = 0, 1, \ldots, d(i)$, $i = 1, 2, \ldots, m$. Then, from the definition of $d(i)$, we see that $\tilde{A}^{\mathrm{T}} x = \tilde{b}$.

Define m_0 to be the number of columns in $\tilde{A}$, so that

$$m_0 = d(1) + d(2) + \cdots + d(m) + m.$$

Then from Theorem 4.10, if x is basic m_0 is at least as large as n. If we were following the same pattern as for the countable semi-infinite program, we could now remark that x being non-degenerate implied that m_0 was exactly n. However things are more complex in this case, and in fact no basic feasible x will be non-degenerate. This is because $D(x)$ only contains functions which are infinitely differentiable at the points s_i, $i = 1, 2, \ldots, m$, and so does not contain all of $C[0, 1]$.

Let $A[0,1]$ be the space of analytic functions on $[0,1]$. Because we have assumed that $a_i(s)$, $i = 1, 2, \ldots, n$ and $b(s)$ are in $A[0,1]$, we may pose SIP2 with the primal constraint space as $A[0,1]$ rather than $C[0,1]$, without altering the set of feasible solutions. After doing this the equality-constrained form of the problem is posed over the space $R^n \times A[0,1]$. With this formulation a basic feasible x is non-degenerate exactly when $m_0 = n$, in which case $\tilde{A}$ is square invertible. Thus if x is a non-degenerate basic solution it is determined from $\tilde{b}$ by

$$x = (\tilde{A}^{-1})^{\mathrm{T}} \tilde{b}.$$

Now consider the reduced cost functional x^*. For (u, v) in $R^n \times A[0,1]$ we have

$$\langle (u, v), x^* \rangle = c^{\mathrm{T}} u - c^{\mathrm{T}} (\tilde{A}^{-1})^{\mathrm{T}} q$$

where q is a vector with elements

$$\frac{d^j}{ds^j}(a(s_i)^T u - v(s_i)), \qquad j = 0,1, \ldots, d(i), \quad i = 1,2, \ldots, m.$$

Thus

$$\langle (u, v), x^* \rangle = c^T(\tilde{A}^{-1})^T \tilde{v}.$$

For x^* to be non-negative we must have $\langle (u, v), x^* \rangle \geq 0$ for all analytic functions $v(s) \geq 0$. However $v^{(j)}(s_i)$ may be positive or negative for $j > 0$, and indeed can be chosen freely. It is only those elements of $\tilde{v}$ given by $v(s_i)$, $i = 1, 2, \ldots, m$ which are non-negative. So x^* is non-negative when the elements of $\tilde{A}^{-1}c$ are all zero, except those which correspond to the $v(s_i)$ in $\tilde{v}$ and these are non-negative.

It will make the notation easier if we now specify that the columns of $\tilde{A}$ occur in the order

$$a(s_i), \qquad i = 1, 2, \ldots, m$$

followed by

$$a^{(j)}(s_i), \qquad j = 1,2, \ldots, d(i), \quad i = 1,2, \ldots, m.$$

Moreover we take the elements of $\tilde{b}$ to be arranged in the same order. Define vectors $\lambda \in R^m$ and $\rho \in R^{n-m}$ by

$$\begin{bmatrix} \lambda \\ \rho \end{bmatrix} = \tilde{A}^{-1}c.$$

Thus a non-degenerate basic feasible solution x is optimal if and only if λ is non-negative and ρ is zero.

Suppose that x is a non-degenerate basic feasible solution which is not optimal. We consider two cases depending on whether some element of λ is negative, or some element of ρ is non-zero. The first of these cases gives rise to a pivot step which is along an extreme ray of the set of feasible perturbations, as our previous examples have been, whilst for the second case we define a pivot step of a different kind.

(I) First suppose that $\lambda_k < 0$, for some $1 \leq k \leq m$. Define a new solution, say x', by

$$x' = (\tilde{A}^{-1})^T(\tilde{b} + \delta e_k)$$

where $\delta > 0$. In other words we make $z(s_k) = \delta$ while leaving the other $z^{(j)}(s_i)$ unchanged. This will be feasible for δ small enough. Now

$$\begin{aligned} c^T x' &= c^T(\tilde{A}^{-1})^T \tilde{b} + \delta c^T (\tilde{A}^{-1})^T e_k \\ &= c^T x + \delta \lambda_k \\ &< c^T x \end{aligned}$$

and so x' is an improvement on x.

To ensure that x' is basic, we wish to choose δ just large enough for a new point to enter constr (x). In fact we must take

$$\delta = \min\{-z(s)/[\tilde{A}^{-1}a(s)]_k\},$$

where the minimization is taken over s for which the denominator is negative. This is equivalent to minimizing

$$\frac{1}{z(s)}[\tilde{A}^{-1}a(s)]_k,$$

and this part of the algorithm will use some standard technique for minimizing a function on a line.

(II) Now suppose that the kth element of ρ is non-zero, where $1 \leqslant k \leqslant n-m$. We shall show how to obtain an improved solution. However, this construction is more complex than for the other simplex extensions that we have so far looked at. It will be helpful to assume to begin with that $d(i) = 1$, $i = 1, 2, \ldots, m$, and also that s_k is neither 0 nor 1. This implies that $n = 2m$.

Consider moving the point s_k to $s_k + \delta$ where δ may be positive or negative. This will determine a particular change in x, and we shall write x' for the new value of x. Now we investigate how this change affects the objective function. After moving s_k to $s_k + \delta$, $\tilde{b}$ becomes $\tilde{b}'$, given by

$$\tilde{b}' = \tilde{b} + (\Delta b) + O(\delta^2),$$

where

$$\Delta b = \begin{bmatrix} \delta b^{(1)}(s_k)e_k \\ \delta b^{(2)}(s_k)e_k \end{bmatrix}.$$

Similarly $\tilde{A}$ becomes $\tilde{A}'$, given by

$$\tilde{A}' = \tilde{A} + (\Delta A) + O(\delta^2),$$

where ΔA is a matrix whose columns are zero except the kth, which is $\delta a^{(1)}(s_k)$, and the $(m+k)$th which is $\delta a^{(2)}(s_k)$. Thus

$$\begin{aligned} c^{\mathrm{T}}x' &= c^{\mathrm{T}}(\tilde{A}'^{-1})^{\mathrm{T}}\tilde{b}' \\ &= c^{\mathrm{T}}(\tilde{A}^{\mathrm{T}} + (\Delta A)^{\mathrm{T}})^{-1}(\tilde{b} + \Delta b)) + O(\delta^2). \end{aligned}$$

Now

$$(\tilde{A}^{\mathrm{T}} + (\Delta A)^{\mathrm{T}})^{-1} = (\tilde{A}^{\mathrm{T}})^{-1} - (\tilde{A}^{\mathrm{T}})^{-1}(\Delta A)^{\mathrm{T}}(\tilde{A}^{\mathrm{T}})^{-1} + O(\delta^2).$$

So

$$\begin{aligned} c^{\mathrm{T}}x' &= c^{\mathrm{T}}(\tilde{A}^{\mathrm{T}})^{-1}\tilde{b} - c^{\mathrm{T}}(\tilde{A}^{\mathrm{T}})^{-1}(\Delta A)^{\mathrm{T}}(\tilde{A}^{\mathrm{T}})^{-1}\tilde{b} + c^{\mathrm{T}}(\tilde{A}^{\mathrm{T}})^{-1}(\Delta b) + O(\delta^2) \\ &= c^{\mathrm{T}}x - (\lambda^{\mathrm{T}}, \rho^{\mathrm{T}})((\Delta A)^{\mathrm{T}}x - (\Delta b)) + O(\delta^2). \end{aligned}$$

But

$$(\Delta A)^{\mathrm{T}}x - (\Delta b) = \begin{bmatrix} \delta(a^{(1)}(s_k)^{\mathrm{T}}x - b^{(1)}(s_k))e_k \\ \delta(a^{(2)}(s_k)^{\mathrm{T}}x - b^{(2)}(s_k))e_k \end{bmatrix}$$

and the top part of this vector is $z^{(1)}(s_k)e_k$ which is zero. Hence

$$\begin{aligned} c^{\mathrm{T}}x' &= c^{\mathrm{T}}x - \delta\rho_k(a^{(2)}(s_k)^{\mathrm{T}}x - b^{(2)}(s_k)) + O(\delta^2) \\ &= c^{\mathrm{T}}x - \delta\rho_k z^{(2)}(s_k) + O(\delta^2). \end{aligned}$$

Now $z^{(2)}(s_k) > 0$, as z remains positive in the region of s_k. Thus we obtain an improvement by moving s_k to $s_k + \delta$ for small enough δ, where δ is chosen to have the same sign as ρ_k.

Things become more complicated if either (a) $s_k = 0$ or 1, or (b) $d(k) > 1$. In case (a) we may have to consider solutions with $z(s_k) = \delta^2$ and $z^{(1)}(s_k) = \delta$. In case (b) we may have to replace s_k by two (or more if $d(k) > 3$) points at $s_k + \delta$ and $s_k + \varepsilon$.

4.5.3 An example to illustrate the primal and dual algorithms

In this section we illustrate the primal and dual algorithms with a very simple example. Consider the following continuous semi-infinite program.

$$\begin{array}{ll} \text{minimize} & x_1 + x_2 \\ \text{subject to} & (1-s)x_1 + sx_2 \geqslant s(1-s), \qquad s \in [0, 1]. \end{array}$$

This has as its dual problem

$$\begin{array}{ll} \text{maximize} & \int_0^1 s(1-s)\,\mathrm{d}w(s) \\ \text{subject to} & \int_0^1 (1-s)\,\mathrm{d}w(s) = 1, \\ & \int_0^1 s\,\mathrm{d}w(s) = 1, \\ & w \geqslant 0, \quad w \in M_r[0, 1]. \end{array}$$

First we apply the primal simplex extension of Section 4.5.2 to this problem. A solution is determined by constr(x). We begin by taking constr$(x) = \{\frac{1}{3}\}$. Now

$$a(s) = \begin{bmatrix} 1-s \\ s \end{bmatrix}, \qquad a'(s) = \begin{bmatrix} -1 \\ 1 \end{bmatrix},$$

$$b(s) = s(1-s), \qquad b'(s) = 1 - 2s, \qquad c = \begin{bmatrix} 1 \\ 1 \end{bmatrix}.$$

To calculate x we take $d(1)$ as 1, so

$$\tilde{A} = \begin{bmatrix} \frac{2}{3} & -1 \\ \frac{1}{3} & 1 \end{bmatrix}, \qquad \tilde{b} = \begin{bmatrix} \frac{2}{9} \\ \frac{1}{3} \end{bmatrix}.$$

Thus

$$\tilde{A}^{-1} = \begin{bmatrix} 1 & 1 \\ -\frac{1}{3} & \frac{2}{3} \end{bmatrix}, \qquad x = (\tilde{A}^{-1})^{\mathrm{T}}\tilde{b} = \begin{bmatrix} \frac{1}{9} \\ \frac{4}{9} \end{bmatrix}.$$

It is easy to check that this is feasible. To see whether it is optimal we calculate the reduced costs λ and ρ. They are given by

$$\begin{bmatrix} \lambda \\ \rho \end{bmatrix} = \tilde{A}^{-1}c = \begin{bmatrix} 2 \\ \frac{1}{3} \end{bmatrix}$$

Thus ρ is non-zero and the current solution is not optimal. As ρ is positive we wish to move the point in constr(x) to the right. How far we move it must be determined by a line search to maximize the objective value. If this is carried out we shall move to a value of $\frac{1}{2}$ for the point in constr(x). Then

$$\tilde{A} = \begin{bmatrix} \frac{1}{2} & -1 \\ \frac{1}{2} & 1 \end{bmatrix}, \quad \tilde{b} = \begin{bmatrix} \frac{1}{4} \\ 0 \end{bmatrix}.$$

Thus

$$\tilde{A}^{-1} = \begin{bmatrix} 1 & 1 \\ -\frac{1}{2} & \frac{1}{2} \end{bmatrix}, \quad x = \begin{bmatrix} \frac{1}{4} \\ \frac{1}{4} \end{bmatrix}.$$

Hence $\lambda = 2$ and $\rho = 0$, so this solution is optimal.

Now we shall apply the dual algorithm of Section 4.5.1 to this problem. We need to start from an initial dual feasible solution, which is determined by supp(w). Take supp(w) to be $\{\frac{1}{3}, \frac{3}{4}\}$. Then

$$\hat{A} = \begin{bmatrix} \frac{2}{3} & \frac{1}{4} \\ \frac{1}{3} & \frac{3}{4} \end{bmatrix}, \quad \hat{b} = \begin{bmatrix} \frac{2}{9} \\ \frac{3}{16} \end{bmatrix}.$$

So

$$\hat{A}^{-1} = \begin{bmatrix} \frac{9}{5} & -\frac{3}{5} \\ -\frac{4}{5} & \frac{8}{5} \end{bmatrix}, \quad \hat{w} = \hat{A}^{-1}c = \begin{bmatrix} \frac{6}{5} \\ \frac{4}{5} \end{bmatrix}.$$

Hence w is feasible. The reduced cost functional w^* is given by

$$\begin{aligned} w^*(s) &= b(s) - a(s)^{\mathrm{T}}\hat{A}^{-1}\hat{b} \\ &= s(1-s) - (1-s)/4 - s/6 \\ &= -(s-\tfrac{1}{3})(s-\tfrac{3}{4}). \end{aligned}$$

This is positive between $\frac{1}{3}$ and $\frac{3}{4}$, so that this choice of w is not optimal. In fact $w^*(s)$ is maximized at

$$s = (\tfrac{1}{3} + \tfrac{3}{4})/2 = \tfrac{13}{24},$$

and this is the point which we choose to introduce into the support of w. Equation (5) in Section 4.5.1 indicates which of the two existing points drops out of the support of w; this turns out to be the point $\frac{3}{4}$. So we obtain a new support $\{\frac{1}{3}, \frac{13}{24}\}$.

Then the whole process can be repeated. At each stage we will introduce a point half way between the existing points; thus successive iterations of the algorithm act like a bisection search for the point $\frac{1}{2}$, which is reached only in the limit.

4.6 THE UNIFORM APPROXIMATION PROBLEM

Some of the most important applications of semi-infinite programming are to approximation problems. We shall look in this section at one such problem, that of uniform approximation. Suppose that we are given a function f, defined on $[0, 1]$ say, and we wish to approximate it by a linear combination of some other given functions $g_1, g_2, \ldots, g_n$ also defined on $[0, 1]$. One objective we might have is to minimize the maximum difference between f and the approximating function, that is to minimize max $|f(s)-p(s)|$, where p is a linear combination of the g_i. This is called uniform approximation or minimax approximation.

It is not difficult to formulate this problem as a semi-infinite linear program. We introduce an extra scalar variable, r, to measure the maximum error. Then the uniform approximation problem takes the form:

$$\text{UAP:}\quad \text{minimize} \quad r$$

$$\text{subject to} \quad \sum_{i=1}^{n} \lambda_i g_i(s) + r \geqslant f(s), \qquad s \in [0,1],$$

$$-\sum_{i=1}^{n} \lambda_i g_i(s) + r \geqslant -f(s), \qquad s \in [0,1].$$

This is a continuous semi-infinite program. The only difference between this form of program and that of SIP2 is that here there are two sets of constraints, both sets indexed by s in $[0,1]$.

We shall assume that the functions $f, g_1, \ldots, g_n$ are continuous. We are now in a position to apply any of the theorems of Section 4.4; the fact that the problem is posed with two sets of constraints makes no essential difference.

From the form of SIP2* it is easy to see that the dual problem to UAP is as follows:

$$\text{UAP*:}\quad \text{maximize} \quad \int_0^1 f(s)\,\mathrm{d}w_1(s) - \int_0^1 f(s)\,\mathrm{d}w_2(s)$$

$$\text{subject to} \quad \int_0^1 g_i(s)\,\mathrm{d}w_1(s) - \int_0^1 g_i(s)\,\mathrm{d}w_2(s) = 0, \qquad \mathrm{i} = 1,2, \ldots, n,$$

$$\int_0^1 \mathrm{d}w_1(s) + \int_0^1 \mathrm{d}w_2(s) = 1,$$

$$w_1 \geqslant 0, \quad w_2 \geqslant 0, \qquad w_1, w_2 \in M_r[0,1].$$

An immediate consequence of Theorem 4.8 is that we need only consider solutions to UAP* which are measures concentrated on a finite number of points. Moreover we can apply Corollary 4.7 to show that, if UAP* has a finite value, then the values of UAP and UAP* coincide (and UAP is solvable, though we do not make this explicit in Theorem 4.4)

There is a well known and very effective procedure for solving the uniform approximation problem that we have posed. It is called the Remez exchange algorithm, and we shall give a brief description of this method before showing how it is related to the methods we described in Section 4.5.

Suppose first of all that the functions g_i, $i = 1,2, \ldots, n$ satisfy the *Haar condition*. This states that for any n distinct points $s_1, s_2, \ldots, s_n$ in $[0,1]$ the $n \times n$ matrix whose (i,j)th element is $g_i(s_j)$ is non-singular. This condition is satisfied for example when the functions g_i are just powers of s, that is $g_i(s) = s^{i-1}$, $s \in [0, 1]$, $i = 1, 2, \ldots, n$.

We write Ω for the space of functions which are linear combinations of the functions $\{g_i\}$. Assuming the Haar condition, it is possible to prove that an optimal choice of the function p in Ω has the property that it reaches its maximum difference from f at exactly $n+1$ points, with this difference taking a positive and negative sign alternately at these $n+1$ points (taken in their natural order along the line). Indeed the optimal p is unique and is the only linear combination of the g_i with this property.

The Remez exchange algorithm works by iteratively updating a set of $n+1$ reference points, $s_0, s_1, \ldots, s_n$ (called the reference set), where $0 \leqslant s_0 < s_1 < \cdots < s_n \leqslant 1$. These reference points are used to determine a function p, given by

$$p(s) = \Sigma \lambda_i g_i(s),$$

which satisfies the $n+1$ equations

$$f(s_i) - p(s_i) = (-1)^i h, \qquad i = 0, 1, \ldots, n. \tag{7}$$

Here h is also determined by the equations (7). The Haar condition ensures that these equations have a unique solution.

Let the error function $e(s)$ be defined by $e(s) = f(s) - p(s)$. The next step is to find the maximum modulus of e in $[0, 1]$. If this is equal to $|h|$, then we have reached the optimum. Suppose that e achieves its maximum modulus at t. The point t is introduced into the reference set and one of the two adjacent reference points, say s_i or s_{i+1}, is dropped from the reference set. As $p(s_i)$ and $p(s_{i+1})$ have opposite signs, there is a unique choice of the point to drop which retains the property that $p(s_i)$ alternates in sign as i goes from 0 to n. Because of the Haar condition this is true generally even if $t < s_0$ or $t > s_n$.

Having generated a new reference set the whole procedure can be repeated. It can be shown that the modulus of h, determined from (7) at each iteration, increases monotonically and that the functions p generated by this procedure converge to the best uniform approximation to f. We have summarized the so-called 'one-point exchange algorithm'; a more efficient algorithm would insert more than one point into the reference set at each iteration.

As we shall see, the Remez exchange algorithm is just the dual simplex extension, described in Section 4.5.1, applied to the uniform approximation problem. Before we show this we need to quote one important theorem, whose proof can be found in Powell (1981).

Theorem 4.11 *Let $\{s_0, s_1, \ldots, s_n\}$ be a reference set and suppose that the Haar condition is satisfied; then p is a function in Ω which minimizes*

$$\max \{|f(s_i) - p(s_i)|, i = 0, 1, \ldots, n\}$$

if and only if p solves (7) for some h.

From this theorem we can make a deduction about the form of basic feasible solutions for UAP*.

Theorem 4.12 *Suppose the Haar condition is satisfied and let the pair* (w_1, w_2) *be a basic feasible solution for* UAP*. *Then* $\text{supp}(w_1)$ *and* $\text{supp}(w_2)$ *contain exactly* $n+1$ *elements in total and are interleaved, in the sense that if* $s_1 < s_2$ *are in* $\text{supp}(w_1)$, *then there is some* t *in* $\text{supp}(w_2)$ *such that* $s_1 < t < s_2$, *and vice versa.*

Proof. As (w_1, w_2) is a basic feasible solution, we can use the result of Section 4.5.1 to see that their combined support contains no more than $n+1$ elements. Suppose now that the combined support contains only n elements. Then the first n constraints of UAP* take the form $L\bar{w} = 0$. Here L is a matrix whose ith row has elements $g_i(s_j)$, where the s_j are the elements of $\text{supp}(w_1)$ and $\text{supp}(w_2)$, and $\bar{w}$ is a vector composed of the n masses in w_1 and w_2 with appropriate signs. From the Haar condition, L is invertible, and this implies that the masses in w_1 and w_2 are all zero. This contradicts the final constraint in UAP*. The argument is easily extended to the case where there are fewer than n elements in the combined support of w_1 and w_2. Hence the first part of the theorem is proved.

Now consider the problem of minimizing

$$\max \{|f(s_i) - p(s_i)| : s_i \in \text{supp}(w_1) \cup \text{supp}(w_2)\}$$

over p in Ω. This can be formulated as a finite linear program:

$$\begin{aligned} &\text{minimize} && r \\ &\text{subject to} && \sum_{i=1}^{n} \lambda_i g_i(s_j) + r \geqslant f(s_j), \\ & && -\sum_{i=1}^{n} \lambda_i g_i(s_j) + r \geqslant -f(s_j), \\ & && \qquad\qquad s_j \in \text{supp}(w_1) \cup \text{supp}(w_2). \end{aligned}$$

It is not difficult to see that a vector whose non-zero elements are the masses which make up w_1 and w_2 is feasible for the dual of this problem. Moreover, this dual solution is complementary slack to the primal solution which satisfies

$$\begin{aligned} \sum_{i=1}^{n} \lambda_i g_i(s_j) + r &\geqslant f(s_j), \qquad && s_j \in \text{supp}(w_1), \\ -\sum_{i=1}^{n} \lambda_i g_i(s_j) + r &\geqslant -f(s_j), \qquad && s_j \in \text{supp}(w_2). \end{aligned} \tag{8}$$

Hence, from a standard complementary slackness result for finite linear programs, the solution to the equations (8) provides the optimal solution to the finite primal problem. But from Theorem 4.11 the unique optimal solution has the interleaved property, hence establishing the second part of the result. □

Now we can explore the form of the dual simplex extension for UAP*. It is clear that a basic solution is determined from the points in $\text{supp}(w_1)$ and $\text{supp}(w_2)$. From Section 4.5.1 we know that (w_1, w_2) is optimal for UAP* if and only if $(\hat{A}^{-1})^{\text{T}}\hat{b}$ is feasible for UAP. Now $\hat{A}$ is the matrix with columns $(g_1(s_i), \ldots g_n(s_i), 1)^{\text{T}}$ for s_i in $\text{supp}(w_1)$ and $(-g_1(s_j), \ldots, -g_n(s_j), 1)^{\text{T}}$ for s_j in supp(w_2); $\hat{b}$ is a vector with elements $f(s_i)$ for s_i in supp(w_1) and $-f(s_j)$ for s_j in supp(w_2). Thus $(\hat{A}^{-1})^{\mathrm{T}}\hat{b}$ is just the solution of the equations (7) when solved for $(\lambda_1, \ldots, \lambda_n, h)$.

This is feasible for UAP if and only if

$$\max \{|f(s) - \Sigma\lambda_i g_i(s)| : s \in [0, 1]\} = h.$$

Otherwise, we make a pivot in the dual simplex extension by introducing a new point into the combined support of w_1 and w_2. This new point was called s_0 in Section 4.5.1; we should choose it at the point where the error function achieves its maximum modulus (to make an improvement, the modulus of the error function at the new point must be larger than h).

We must also drop one point from the combined support of w_1 and w_2. This point is determined from equation (1) of Section 4.5.1. It might seem that this could lead to any of the support points being dropped. However, as we are moving to a new basic feasible solution, the point to be dropped is determined uniquely from the interleaved property of Theorem 4.12.

4.7 NOTES

Semi-infinite linear programs were first introduced by Charnes *et al* (1963). Their paper dealt with countable semi-infinite programs and they endeavoured to prove a strong duality result by building on some work of Haar(1924). However, their paper contained an error which was later corrected by Duffin and Karlovitz (1965). Duffin and Karlovitz also introduced the idea of truncated problems (dealt with briefly in Section 4.2).

In the papers of Charnes *et al* (1963, 1965) and Duffin and Karlovitz (1965) the idea of canonical closure plays a prominent role. The set of constraints for the semi-infinite program is called canonically closed if there exist some constants $\lambda(s)$, $s \in S$, such that the set $\{\lambda(s)\ (a(s), b(s)) : s \in S\}$ is compact in R^{n+1}. This condition, together with an interior point condition (as in Theorem 4.5), is enough to ensure strong duality, independently of the nature of the set S. For a proof of this result see Glashoff(1979).

More recent work on duality theory for semi-infinite linear programs is described in the papers of Karney (1981) (from which the example of 4.2.1 is taken), Borwein (1981, 1983a) and Duffin, Jeroslow and Karlovitz(1983).

There is a close connection between the theory of semi-infinite linear programming and the theory of moment problems. Briefly stated, moment problems are concerned either with establishing the values that are possible for a vector with components $E(f_1(X)), E(f_2(X)), \ldots, E(f_n(X))$, when X is a random variable taking values in some given subset of Euclidean space, or with finding bounds on $E(f_0(X))$ given $E(f_1(X)), E(f_2(X)), \ldots, E(f_n(X))$. These problems are very closely related to the dual semi-infinite linear problem. Theorems 4.6 and 4.8, for example, are fairly standard results from moment theory (Rogosinski(1958) and Isii (1960)). The papers by Duffin (1967) and Gustafson (1970) treat particular kinds of moment problem as semi-infinite linear programs, and the connections between the two fields are further discussed by Glashoff (1979) and Kemperman (1983). Karr (1983) discusses the nature of the extreme points for certain moment problems.

The paper by Gustafson (1970) is concerned with methods for the numerical solution of semi-infinite programs. Other papers on the same topic are those of Gustafson and Kortanek (1973), Gribik (1979), Gustafson (1983), Hettich (1983), Coope and Watson (1985) and Watson (1985). These papers all concentrate on algorithms for the dual problem and for the most part relate them to cutting plane methods, rather than to simplex extensions. Our rather different viewpoint has led us to adopt a different approach in this chapter. The book by Glashoff and Gustafson (1983) develops the theory of semi-infinite programming from first principles. Another book which treats the same area is that by Krabs (1979), though he includes a great deal of other material on convex and non-linear optimization and approximation problems.

It has been recognized for a long time that many problems in approximation theory can be posed as semi-infinite programs. An early paper on the applications of linear programming duality theory to approximation problems is that by Duffin and Karlovitz (1968). An excellent treatment of the uniform approximation problem can be found in the book by Powell (1981). The connection between the Remez exchange algorithm and the simplex method for the dual problem is well known and was first made by Stiefel (1960) in a finite-dimensional setting. The papers by Watson (1975) and Hoffman and Klostermair (1976) both describe methods for solving more general uniform approximation problems using techniques from semi-infinite linear programming.

Another class of problem related to semi-infinite programs are *separably infinite* programs. These problems have an infinite number of variables and an infinite number of constraints. However, only a finite number of constraints have an infinite number of variables, and only a finite number of variables appear in an infinite number of constraints. Separably infinite programs have been studied by Charnes, Gribik and Kortanek (1980) and Borwein and Kortanek (1983).

Problems

4.1 Prove the assertion in the proof of Theorem 4.2, that if c is in the interior of M_A we may choose $n+1$ points in M_A with c in the interior of their convex hull.

4.2 Consider the problem P1* of Section 4.2.1. Show that the optimal solution is a degenerate basic solution and identify the reduced cost functional.

4.3 Find M_A and M_B for the example given at the end of Section 4.4 and check that none of the conditions of Theorems 4.4, 4.5 or 4.7 apply to it.

4.4 Carefully justify the characterization given in Section 4.5.2 of non-negative x^* in terms of the elements of $\tilde{A}^{-1}c$.

4.5 Calculate the change in the objective function for SIP2 if $d(k) = 3$ and s_k is changed to two points, one at $s_k + \delta$ and the other at $s_k + \varepsilon$. Show that if the kth element of ρ is non-zero, then ε and δ can be chosen to make an improvement in the objective function.

4.6 (Separably infinite program) Consider the following separably infinite program. Find x in R^n and μ a measure on $[0,1]$ to

$$\begin{aligned} &\text{minimize} && c^{\mathrm{T}}x + \int_0^1 f(s)\,\mathrm{d}\mu(s) \\ &\text{subject to} && a(s)^{\mathrm{T}}x \geqslant b(s), \quad s \in [0,1], \\ & && Ax + \int_0^1 g(s)\,\mathrm{d}\mu(s) = d, \\ & && \mu \geqslant 0, \end{aligned}$$

where $g(s)$ and d are m-vectors and A is an $m \times n$ matrix. Assuming that a and b are analytic functions and that a feasible solution (x, μ) is given in which μ has a finite number of mass points, find a matrix whose rows are linearly independent if and only if this solution is basic.

4.7 Consider a Markov chain with states $1, 2, \ldots, n$. In any state actions can be made which are indexed by the set $[0,1]$. If in state i action s is made, then a cost $c_i(s) > 0$ is incurred and the system moves to state k with probability $p_i(s, k)$. Suppose that all costs are discounted at a rate α. Then the optimal value function $V(i)$ is given by the solution to the dynamic programming equation

$$V(i) = \inf_{s \in [0,1]} \{c_i(s) + \alpha \sum_{j=1}^{n} p_i(s,j)\, V(j)\}.$$

Show that the optimal value function can be obtained as the solution to the semi-infinite linear program

$$\begin{aligned} &\text{maximize} && \sum_{i=1}^{n} x_i \\ &\text{subject to} && x_i - \alpha \sum_{j=1}^{n} p_i(s,j) x_j \leqslant c_i(s), \\ & && i = 1, 2, \ldots, n, \; s \in [0,1]. \end{aligned}$$

Find the dual of this program.

4.8 Show that the algebraic dual of SIP1 is not SIP1*.

4.9 Find an example of SIP1 with a basic feasible solution having constr $(x) = \emptyset$.

5

The Mass-transfer Problem

5.1 INTRODUCTION

We come now to fully infinite linear programs, in which both the variable and constraint spaces are of infinite dimension. Our approach will be to consider analogues and extensions of finite-dimensional problems, and in particular those with special structure. We might expect to make more progress with highly structured problems, particularly in attempting to formulate algorithms. In later chapters, we shall look at problems associated with optimizing flows in networks, where the structure of the constraints is of a special form determined by the characteristics of the network, and at a class of continuous-time optimal control problems with state positivity constraints. Our aim in this chapter is to investigate an infinite-dimensional extension of the classical transportation problem, called the mass-transfer problem.

Suppose we wish for some reason to level a plot of land. Then we will take earth from all those places where the level is above the mean, and redistribute it in the hollows where the level is below the mean. To minimize the work involved, we will wish to find a pattern of redistribution that minimizes the overall movement of earth. A similar problem arises in the construction of any earthwork, and not surprisingly its study has a long history. In fact, systematic mathematical study of the problem seems to have begun in eighteenth-century France, where the problem of 'déblais et remblais' (cutting and filling) was important for the efficient construction of fortifications.

The analogy between these problems and the transportation problem is apparent. The latter requires that a number of discrete requirements be met from a set of discrete sources in a way that minimizes the transportation cost. In the cutting and filling problem, the sources are the points from which earth is to be removed, and the destinations are the points where filling is needed. We will formulate a general version of the problem, capable of incorporating both the finite and infinite cases, called the continuous transportation problem.

5.2 THE CONTINUOUS TRANSPORTATION PROBLEM

The classical transportation problem is a linear program posed in R^{mn} as follows:

$$\text{TP:}\quad \text{minimize} \quad \sum_{i=1}^{m}\sum_{j=1}^{n} c_{ij}x_{ij}$$

$$\text{subject to} \quad \sum_{j=1}^{n} x_{ij} = a_i, \qquad i = 1,2, \ldots, m,$$

$$\sum_{i=1}^{m} x_{ij} = b_j, \qquad j = 1,2, \ldots, n,$$

$$x_{ij} \geqslant 0, \qquad i = 1,2, \ldots, m, \quad j = 1,2, \ldots, n.$$

The decision variables x_{ij} represent the amounts shipped from source i to destination j. The demand at destination j is b_j and the supply at source i is a_i. In order for the problem to be consistent, the total supply must equal the total demand. If it is less, then the demands cannot all be met. If total supply exceeds total demand, the problem may be reformulated with a dummy destination to take up the excess. Actually, there is no theoretical necessity to treat the supply constraint like this, but putting the problem in the exact form given allows it to be elegantly solved by the transportation version of the simplex algorithm (see, for example, Dantzig (1963)).

The continuous transportation problem which we now pose is an abstraction of TP in which the distribution of supply and demand is represented by measures on two compact spaces U and V, and the assignment of supplies to meet demands is represented by a measure on $U \times V$. For any measure ρ on $U \times V$, we denote by $P_U\rho$ the projection of ρ onto U. Thus for any measurable set $S \subset U$, $(P_U\rho)(S) = \rho(S \times V)$. We define the projection P_V onto V similarly. The continuous transportation problem is

$$\text{CTP:}\quad \text{minimize} \quad \int_{U \times V} c(u, v)\,\mathrm{d}\rho(u, v)$$

$$\text{subject to} \quad P_U\rho = \mu_1,$$

$$P_V\rho = \mu_2,$$

$$\rho \geqslant 0.$$

Hence c is a continuous function on $U \times V$, and ρ, μ_1, μ_2 are regular Borel measures on $U \times V$, U, and V respectively. Since ρ is constrained to be non-negative, the problem is inconsistent unless μ_1 and μ_2 are also non-negative, and for consistency we must also have $\mu_1(U) = \mu_2(V)$. When CTP is viewed as a mass-transfer problem, ρ represents a translocation of mass between U and V, and the constraints define the initial distribution of mass in U and the required distribution of mass in V.

5.2.1 The dual problem

Let F be any compact space. As in Chapter 4, we denote by $M_r(F)$ the space of regular, signed Borel measures on F. We write $C(F)$ for the space of continuous real-valued functions on F, with associated norm defined by

$$\|f\| = \max_{x \in F} \{|f(x)|\}.$$

By the Riesz representation theorem, we can identify $M_r(F)$ with the topological dual of $C(F)$, with $\langle f, \mu \rangle = \int f \, d\mu$ for each $f \in C(F)$ and $\mu \in M_r(F)$. This suggests that we pose a dual problem for CTP in $C(U) \times C(V)$. In the paired space formulation of Chapter 3, we adopt the following scheme:

$$\text{CTP:} \quad \begin{bmatrix} M_r(U \times V) \\ C(U \times V) \end{bmatrix} \begin{matrix} \rightarrow \\ \leftarrow \end{matrix} \begin{bmatrix} M_r(U) \times M_r(V) \\ C(U) \times C(V) \end{bmatrix} : \text{CTP*}$$

To formulate the dual, we must work out how to represent the adjoint of the primal constraint map with this choice of spaces. Suppose that $(r, s) \in C(U) \times C(V)$; then

$$\langle (P_U \rho, P_V \rho), (r, s) \rangle = \int_U r(u) \, (dP_U \rho)(u) + \int_V s(v) \, (dP_V \rho)(v),$$

$$= \int_{U \times V} \{\bar{r}(u, v) + \bar{s}(u, v)\} \, d\rho(u, v),$$

where $\bar{r}(u, v) = r(u)$, $\bar{s}(u, v) = s(v)$ for all $(u, v) \in U \times V$. We deduce that the image of (r, s) under the adjoint of the primal constraint map is $\bar{r} + \bar{s}$, and arrive at the dual problem

$$\begin{aligned} \text{CTP*:} \quad & \text{maximize} && \int_U r(u) \, d\mu_1(u) + \int_V s(v) \, d\mu_2(v) \\ & \text{subject to} && r(u) + s(v) \leqslant c(u, v), \quad (u, v) \in U \times V, \end{aligned}$$

where r and s are continuous functions on U and V respectively. The standard weak duality theorem becomes

Lemma 5.1 *If ρ is feasible for* CTP, *and (r, s) is feasible for* CTP*, *then*

$$\int_{U \times V} c(u, v) \, d\rho(u, v) \geqslant \int_U r(u) \, d\mu_1(u) + \int_V s(v) \, d\mu_2(v),$$

with equality if and only if ρ is optimal for CTP *and (r,s) is optimal for* CTP*.

Having set up primal and dual problems, we naturally ask what sort of duality holds between them. Ideally, our chosen setting will be such that CTP and CTP* are strongly dual (both solvable, with no duality gap).

5.2.2 Solvability and duality

We should like to establish three things: that CTP is solvable, that there is no duality gap between CTP and CTP*, and that CTP* is also solvable. The first of these is straightforward, since $M_r(U \times V)$ is the topological dual of $C(U \times V)$.

Thus Theorem 3.20 can be used to guarantee solvability of CTP. If the value of CTP* can be shown to be finite, then Theorem 3.13 ensures that there is no duality gap. Under mild conditions, this too is simple enough. Solvability of CTP* presents a problem, however. The space of continuous functions on a compact set is not the dual of any normed space, so questions of solvability for problems posed in such spaces cannot be settled by using Alaoglu's theorem. There being no other tool of general applicability with which to attack the problem, it is inevitable that methods specific to this problem must be used.

A natural line of attack is to re-pose the problem in a different space, with the compactness properties that one needs, then use conditions on the problem data to show that an optimal solution in the enlarged space must necessarily lie in the original space. This approach is often useful, and will be used more than once in the remainder of this book. In the present case, under mild conditions, we can demonstrate solvability for CTP* in this way. We prove this, and the other two results, in the theorem below. This theorem makes use of the fact, assumed in formulating CTP, that c is a continuous function.

Theorem 5.2 *Assume that U and V are compact subsets of a linear topological vector space. Then*

(*a*) CTP *is solvable;*
(*b*) *there is no duality gap for* CTP *and* CTP*;
(c) CTP* *is solvable.*

Proof. (a) We begin by showing that CTP is consistent. We have already assumed that $\mu_1(U) = \mu_2(V)$. Denote their common value by m. Let μ_3 be the product measure defined from μ_1, μ_2 on $U \times V$. That is, for each continuous function $g \in C(U \times V)$,

$$\int_{U \times V} g(u, v)\, d\mu_3(u, v) = \int_U \left\{ \int_V g(u, v)\, d\mu_2(v) \right\} d\mu_1(u).$$

Then it is easily verified that μ_3 is feasible for CTP. The feasible set for CTP is clearly bounded, and the constraint map and objective functional are weak* continuous, so the result follows from Theorem 3.20.

(b) CTP* is consistent, since c is bounded below, and its value is finite since CTP is solvable. Hence by Theorem 3.13, there is no duality gap.

(c) We start by re-posing CTP* with $r \in L_\infty(\mu_1)$, $s \in L_\infty(\mu_2)$, and taking the constraint space as $L_\infty(\mu_3)$, where μ_3 is as defined in (a). In this new setting, CTP* is solvable by Theorem 3.20 provided its feasible set is bounded. This is not true as the problem stands, since if (r, s) is feasible, we may add an arbitrary constant function to r and subtract it from s, and the result will again be feasible. This is easily overcome, however, by adding the constraint $r \geqslant \theta$. The resulting constraints are still consistent, and the corresponding feasible set is bounded, and Theorem 3.20 may now be applied.

Having shown that CTP* is solvable in L_∞, we complete the proof by showing that continuity of c implies that if (r, s) is an optimal solution in $L_\infty(\mu_1) \times L_\infty(\mu_2)$, then r and s are continuous. First, observe that if (r, s) is optimal, then defining

$$r_1(u) = \inf_{v \in V} \{c(u, v) - s(v)\}, \qquad u \in U,$$

yields a new optimal solution (r_1, s). (Strictly, s is an equivalence class of functions, all μ_2-essentially equal; in the definition of r_1, s must be taken as any representative of this class.) It is clear that r_1 is feasible; that it is optimal follows from the fact that $r_1 \geqslant r$, and μ_1 is positive. We show that r_1 is continuous. Suppose that $\{v_i\}$ is a sequence in V such that $c(u, v_i) - s(v_i)$ approaches $r_1(u)$. Since $U \times V$ is compact, c is uniformly continuous. Thus for $\varepsilon > 0$, there is some neighbourhood N of θ such that

$$|(c(u, v_i) - s(v_i)) - (c(u', v_i) - s(v_i))| < \varepsilon, \qquad i = 1,2, \ldots$$

whenever $u - u' \in N$, $u, u' \in U$. Since $r_1(u') \leqslant c(u', v_i) - s(v_i)$ by definition, we have

$$r_1(u') \leqslant r_1(u) + \varepsilon$$

for any such u'. Equally, we must have

$$r_1(u) \leqslant r_1(u') + \varepsilon.$$

Hence r_1 is continuous. If we now define

$$s_1(v) = \inf_{u \in U} \{c(u, v) - r_1(v)\}, \qquad v \in V,$$

s_1 will also be continuous, and (r_1, s_1) will be an optimal solution to CTP* in $C(U) \times C(V)$. □

The assumption about the form of U and V is used here only in (c), in order to ensure that the underlying topologies on those sets allow of the concept of uniform continuity. This is possible more generally (see, for example, Schaefer (1971)).

Having shown that CTP and CTP* are strongly dual in a setting where the solutions to both are readily characterized, we proceed to examine the possibility of constructing an algorithm to solve examples of CTP. This requires that we specialize to a more concrete version of the problem, and we do this in the next section, where we begin examining the structure of basic solutions.

5.3 BASIC SOLUTIONS AND PIECEWISE CONTINUOUS ASSIGNMENTS

We turn now to a particular example of CTP, for which $U = V = [0, 1]$, and μ_1 and μ_2 are both Lebesgue measure. Any problem in which μ_1 and μ_2 are absolutely continuous measures on intervals of the real line can be transformed into this form (Problem 5.2). We also assume that c has continuous first partial derivatives. We continue to refer to the problem as CTP.

The existence of a basic optimal solution for CTP is ensured by the weak* compactness of the feasible set. We begin by examining the structure of the basic solutions. Let ρ be a basic feasible solution for CTP. The Lebesgue decomposition theorem says that ρ may be decomposed into three components, $\rho = \rho_1 + \rho_2 + \rho_3$, with ρ_1 absolutely continuous with respect to Lebesgue measure, ρ_2 concentrated on a set of measure zero, and ρ_3 concentrated on a countable set of points. Since ρ is feasible, its projections onto U and V are both equal to Lebesgue measure. Hence ρ_3 must be zero, since otherwise its projections would give finite mass to some single point.

We can also readily see that if ρ is a basic solution for CTP, then ρ_1 cannot be strictly positive on any open set. More precisely, if ρ_1 is defined by f_1, so that

$$\int_{U \times V} g(u, v)\, \mathrm{d}\rho_1(u, v) = \int_{U \times V} g(u, v) f_1(u, v)\, \mathrm{d}\mu(u, v), \qquad \text{for all } g \in C(U \times V),$$

where μ is Lebesgue measure, then for any $\varepsilon > 0$, the set $\{(u, v) : f_1(u, v) \geqslant \varepsilon\}$ has no interior. Otherwise, we can find $\varepsilon > 0$ and a square $(u-\delta, u+\delta) \times (v-\delta, v+\delta)$, $\delta > 0$ such that the value of f_1 exceeds ε throughout this square. We can divide the square into four equal squares, and define a function f to be alternately $+\varepsilon$ and $-\varepsilon$ on these four squares. Then the absolutely continuous measure defined by f projects onto the zero measure in both directions, and clearly $f_1 + f$ and $f_1 - f$ are both non-negative, and so ρ is not basic.

To characterize further an arbitrary basic solution is difficult, but if ρ is optimal, we can use the properties of its complementary dual solution to provide more information about ρ.

Theorem 5.3 *Suppose that $c_{uv} > 0$ on $U \times V$. Then the optimal solution ρ is unique, and*

$$\mathit{supp}\,(\rho) = \{(u, v) : 0 \leqslant u \leqslant 1,\ 0 \leqslant v \leqslant 1,\ u + v = 1\}.$$

Proof. We begin by showing that for each $(u, v) \in \operatorname{supp}(\rho)$, there is no point of $\operatorname{supp}(\rho)$ in the positive orthant with vertex (u, v). We do this by considering an optimal solution to the dual problem, which must be complementary to ρ, denoted by (r, s). Thus

$$r(u) + s(v) \leqslant c(u, v), \qquad (u, v) \in U \times V,$$

with equality on the support of ρ. Suppose (u, v) and (u_1, v_1) are two points in $\operatorname{supp}(\rho)$. Then we have, since (r, s) is feasible,

$$c(u_1, v) - r(u_1) - s(v) \geqslant 0,$$
$$c(u, v_1) - r(u) - s(v_1) \geqslant 0.$$

Moreover,

$$\begin{aligned} c(u_1, v_1) - c(u, v) &= (c(u, v_1) - c(u, v)) + (c(u_1, v_1) - c(u, v_1)) \\ &= (c(u, v_1) - c(u, v)) + \int_u^{u_1} c_u(u', v_1)\, \mathrm{d}u' \end{aligned}$$

$$= (c(u, v_1) - c(u, v)) + \int_u^{u_1} \left\{ c_u(u', v) + \int_v^{v_1} c_{uv}(u', v')\,dv' \right\} du'$$

$$= (c(u, v_1) - c(u, v)) + (c(u_1, v) - c(u, v))$$

$$+ \int_u^{u_1} \int_v^{v_1} c_{uv}(u', v')\,dv'\,du'.$$

Combining this equation with the two inequalities yields

$$(c(u_1, v_1) - r(u_1) - s(v_1)) + (c(u, v) - r(u) - s(v)) \geqslant \int_u^{u_1} \int_v^{v_1} c_{uv}(u', v')\,du'\,dv'.$$

The left-hand side of this inequality is zero, which implies that $(u_1 - u)$ and $(v_1 - v)$ must be of opposite sign. Hence there is no point of $\text{supp}\,(\rho)$ in the positive orthant with vertex (u, v). This in turn implies that there is also no point of $\text{supp}\,(\rho)$ in the negative orthant with vertex (u, v).

Now suppose that (u, v) is a point of $\text{supp}\,(\rho)$ not on the given line segment. Then by the theorem $\text{supp}\,(\rho)$ is confined to the sets

$$\Gamma_1 = [u, 1] \times [0, v], \qquad \Gamma_2 = [0, u] \times [v, 1].$$

Neither of these is a square, but as ρ is feasible, $\rho(\Gamma_i)$ must be simultaneously equal to both dimensions of Γ_i, a contradiction. □

A corresponding result holds if c_{uv} is negative. It can be shown from this theorem that on any rectangle where c_{uv} is strictly positive (negative) the support of ρ is confined to the graph of a monotonic decreasing (increasing) function.

These results lead us to look at a special class of basic feasible solutions, for which $\text{supp}\,(\rho)$ is confined to a finite set of disjoint squares whose projections onto each of U and V do not overlap, and in each of which it consists of one or other diagonal of the square. Such a solution is called a *piecewise continuous assignment*. Its support is the closure of the graph of a one-to-one function from [0, 1] to [0, 1]. More generally, we shall say that a function $f: [0, 1] \to [0, 1]$ is a defining function for a basic feasible solution ρ if $\text{supp}\,(\rho)$ is contained in the closure of the graph of f (note that this implies that f is surjective). The function f is usually just the explicit form of the implicit relationship $c(u, v) - r(u) - s(v) = 0$. Any solution with a defining function which is a bijection is called an *assignment*.

Obviously, not every basic feasible solution is a piecewise continuous assignment, and, for an arbitrary problem, we may not be able to find an optimal solution which is an assignment. This contrasts with the finite-dimensional problem, where the condition that each source has unit supply and each destination unit demand implies that the amount shipped from any source to any destination will be either 0 or 1. Figure 5.1 illustrates an example where, though the cost function is perfectly well-behaved, the optimal solution is not a piecewise continuous assignment. In this example $c(u, v) = 4u^2 v - uv^2$, whose contours are

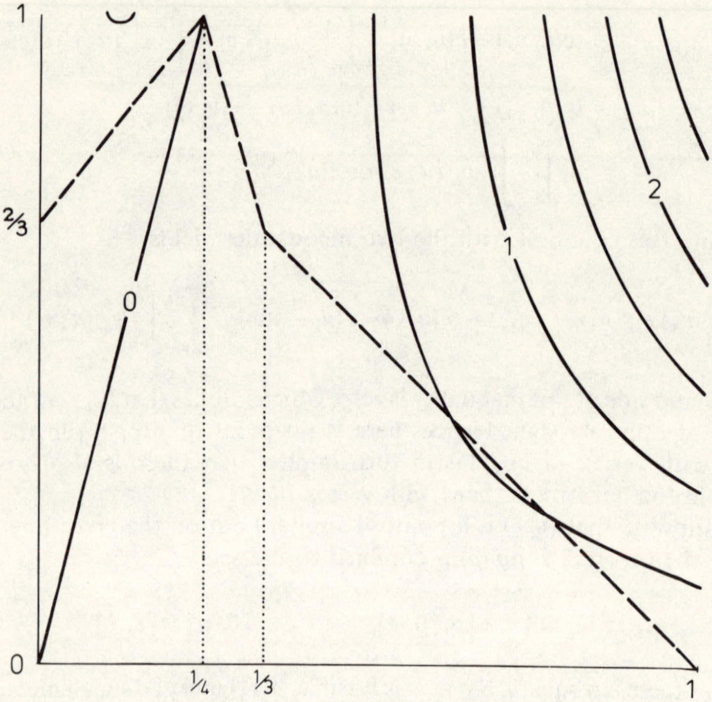

Figure 5.1 An optimal solution which is not a piecewise continuous assignment

shown. The unique optimal solution has the following defining function:

$$\begin{aligned} f(u) &= \tfrac{2}{3} + 4u/3, & 0 \leqslant u < \tfrac{1}{4}, \\ &= 2 - 4u, & \tfrac{1}{4} \leqslant u < \tfrac{1}{3}, \\ &= 1 - u, & \tfrac{1}{3} \leqslant u \leqslant 1. \end{aligned}$$

The graph of this function is the dashed line in the figure.

Although we cannot be certain that an optimal solution can be found which is a piecewise continuous assignment, it is easy to see that an optimal solution can be approximated in value as closely as we please by such a solution. Given any optimal solution ρ, we discretize the square into smaller squares, and approximate ρ in the obvious way by a measure whose support is confined to the diagonals of these squares. This approximation corresponds to some feasible solution of a finite transportation problem, and it is either basic or it can be transformed into a basic solution with improved objective function value. Choosing a sufficiently fine discretization gives the required approximation. This argument can be made rigorous, but in the next section we describe an algorithm which works by producing a sequence of piecewise continuous assignments converging in value to the optimum. This proves the result in a different way.

5.4 AN ALGORITHM FOR THE CONTINUOUS TRANSPORTATION PROBLEM

In looking for a method of solution for the CTP, it is natural to consider discretizing the problem and then solving it using one of the methods available for the standard finite transportation problem. Certainly the value of the optimal solution obtained in this way will approach the value for the continuous problem as the discretization becomes finer. The method is straightforward and robust, but it suffers from two major disadvantages. First, the computation times involved can be very large; though the extent of this problem will depend on how accurate an answer is required and hence on how fine a discretization is made. Secondly, the problem may have an optimal solution of a simple form which is obscured in the discretized version.

For these reasons it is worthwhile to look for a solution algorithm which treats the problem directly. The algorithm we describe below starts from a piecewise continuous assignment and tests for its optimality. If it is not optimal, then a new piecewise continuous assignment is produced which has a strictly improved value for the objective function. This whole process is repeated until an optimal solution is reached. This algorithm is a simplex extension, and in some ways similar to the algorithms we met in Chapter 4. However, there are two complicating factors: first, the basic solutions for CTP are always degenerate; second, the algorithm works entirely with a subclass of the basic solutions, i.e. piecewise continuous assignments.

Computational experience with this algorithm suggests that when the optimal solution is a piecewise continuous assignment, the algorithm will reach exactly this solution in a small number of steps. We shall prove in Section 5.4.1 that even when the solution is not a piecewise continuous assignment the algorithm will converge to the optimal solution.

Before describing the algorithm it is convenient to give a result which bounds the difference between the objective-function value at a trial solution and the optimal value. This result is a special case of the result in Problem 2.4. It will be used again when we come to consider convergence of the algorithm.

Theorem 5.4 *If f is the defining function for a piecewise continuous assignment and there are piecewise continuous functions r and s on $[0, 1]$, such that*

$$r(u) + s(v) \leqslant c(u, v) + \delta, \qquad u, v \in [0,1], \tag{1}$$

$$r(u) + s(f(u)) = c(u, f(u)), \qquad u \in [0,1], \tag{2}$$

then

$$\int_0^1 c(u, f(u))\,\mathrm{d}u - \delta \leqslant V(\mathrm{CTP}) \leqslant \int_0^1 c(u, f(u))\,\mathrm{d}u.$$

Proof. If ρ is the feasible solution corresponding to f, then

$$\int_0^1 c(u, f(u))\,\mathrm{d}u = \int c(u, v)\,\mathrm{d}\rho(u, v),$$

and, as ρ is feasible for CTP, the right-hand inequality is immediate. To prove the left-hand inequality observe that, if $\bar{\rho}$ is any feasible solution for CTP, then

$$\begin{aligned}\int c(u, v)\,\mathrm{d}\bar{\rho}(u, v) &\geqslant \int \{r(u)+s(v)-\delta\}\,\mathrm{d}\bar{\rho}\\ &= \int_0^1 r(u)\,\mathrm{d}u + \int_0^1 s(v)\,\mathrm{d}v - \delta\\ &= \int_0^1 \{r(u)+s(f(u))\}\,\mathrm{d}u - \delta\\ &= \int_0^1 c(u, f(u))\,\mathrm{d}u - \delta. \qquad \square\end{aligned}$$

Suppose now that ρ is a piecewise continuous assignment with describing function f. We first of all construct a particular choice of functions r and s which are complementary slack with ρ (i.e. they satisfy (2)). As in the discrete version of the transportation algorithm these are used to test for optimality, and in the event that the current solution is not optimal, they indicate how to improve it. Considered as a solution to the transportation problem a piecewise continuous assignment is highly degenerate. In the discrete case we would break the degeneracy by including certain of the x_{ij} variables in the basis even though they were zero. The choice of zero-valued x_{ij} to include determines the exact form of r and s, and hence the pivot operation. Similarly, for the continuous problem, different choices of the functions r and s are possible and will lead to different forms of pivot operation.

We define a continuous function r on $[0, 1]$ as follows. Set $r(0) = 0$ and when f is continuous at u with $f(u) = v$, let

$$\mathrm{d}r(u)/\mathrm{d}u = \partial c(u, v)/\partial u.$$

At discontinuities of f let r be determined by continuity. This defines the function r uniquely. We define s from r by

$$s(f(u)) = c(u, f(u)) - r(u), \qquad u \in [0, 1].$$

The functions r and s are called the *reduced-cost functions* for f.

These reduced-cost functions are related to the reduced-cost functional defined in Chapter 2. In fact, writing ρ^* for the reduced-cost functional for ρ, we have

$$\langle \nu, \rho^* \rangle = \int \{c(u, v) - r(u) - s(v)\}\,\mathrm{d}\nu(u, v),$$

for any measure ν in $D(\rho)$. To show that this is the case, let $\nu = \nu_1 + \nu_2$, with $\nu_1 \in B(\rho)$ and $P_U\nu_2 = P_V\nu_2 = 0$. Then

$$\int \{c(u, v) - r(u) - s(v)\}\,\mathrm{d}\nu = \int c(u, v)\,\mathrm{d}\nu - \int \{r(u)+s(v)\}\,\mathrm{d}\nu_1 - \int \{r(u)+s(v)\}\,\mathrm{d}\nu_2$$

$$= \int c(u, v)\,\mathrm{d}\nu - \int c(u, v)\,\mathrm{d}\nu_1 - \int_0^1 r(u)\mathrm{d}P_U\nu_2 - \int_0^1 s(v)\,\mathrm{d}P_V\nu_2$$

$$= \int c(u, v)\,\mathrm{d}\nu - \int c(u, v)\,\mathrm{d}\nu_1$$

$$= \langle P_N\nu, c^* \rangle = \langle \nu, \rho^* \rangle.$$

As ρ is degenerate, ρ^* is defined only on $D(\rho)$ and not on the whole of $M_r(U \times V)$.

We begin by testing for the optimality of ρ. If $r(u) + s(v) \leqslant c(u, v)$ everywhere on $[0,1] \times [0,1]$, then we may set $\delta = 0$ in Theorem 5.4 to show that ρ is optimal for CTP. This can also be deduced from the characterization of the reduced-cost functional as $c(u, v) - r(u) - s(v)$. In fact the functions r and s are also optimal for CTP*; this follows from observing that the reduced-cost function s is a continuous function when the optimal solution to CTP is reached (see Problem 5.3).

Now suppose that $r(u) + s(v)$ is not everywhere less than $c(u, v)$. Then for some choice of u^* and v^*

$$r(u^*) + s(v^*) - c(u^*, v^*) = \delta > 0. \tag{3}$$

Let $z = f^{-1}(v^*)$. By perturbing u^* if necessary, we may choose u^*, v^* so that (3) holds and, in addition, f is continuous at z.

Suppose first that $z > u^*$. We define f' from f as follows:

$$\begin{aligned} f'(u) &= f(z + u^* - u), && u \in (u^*, u^* + \varepsilon),\\ &= f(u - \varepsilon), && u \in (u^* + \varepsilon, z),\\ &= f(u), && \text{otherwise,} \end{aligned}$$

where $\varepsilon > 0$ is chosen smaller than $z - u^*$, and small enough for f to be continuous on the whole of $(z - \varepsilon, z)$. This defines a new piecewise continuous assignment and an example of the change that this produces is shown in Figure 5.2. f' is not defined at some of its discontinuities, but this will not matter in what follows. Indeed we will assume throughout our discussion that a piecewise continuous assignment is defined arbitrarily at its discontinuities.

In the case that $z < u^*$, define f' as follows:

$$\begin{aligned} f'(u) &= f(z + u^* - u), && u \in (u^* - \varepsilon, u^*),\\ &= f(u + \varepsilon), && u \in (z, u^* - \varepsilon),\\ &= f(u), && \text{otherwise.} \end{aligned}$$

Here, as before, $\varepsilon > 0$ must be less than $u^* - z$, and small enough for f to be continuous on the whole of $(z, z + \varepsilon)$.

In the theorem below we show that for ε small enough f' has a strictly improved

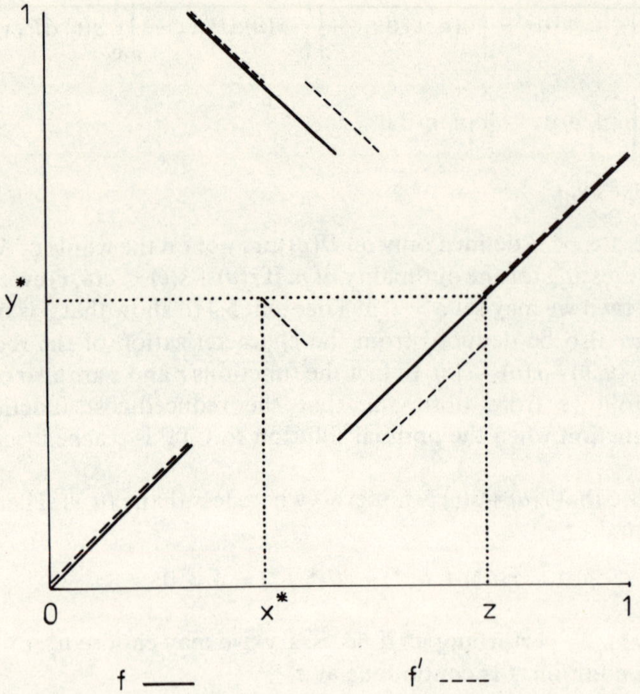

Figure 5.2 One step of the algorithm for CTP

objective-function value compared with f. Note that when we come to prove the convergence result in Section 5.4.1 we will need to modify the definition of f' in the case where f has discontinuities close to z. We begin by introducing a bound on the problem parameters.

As c has continuous first partial derivatives we may choose some number M such that $\partial c/\partial u$ and $\partial c/\partial v$ are bounded in modulus by M. Moreover, we shall assume that $\partial^2 c/\partial u^2$ exists and is bounded in modulus by M.

Theorem 5.5 *If f' is constructed from f as described above, then*

$$\int_0^1 c(u, f(u))\,du - \int_0^1 c(u, f'(u))\,du > \varepsilon\delta - 4\varepsilon^2 M,$$

which is positive for ε chosen small enough.

Proof. When $z > u^*$, we have

$$\int_0^1 c(u, f(u))\,du - \int_0^1 c(u, f'(u)\,du$$

$$= \int_{u^*}^{z-\varepsilon} \{c(u, f(u)) - c(u+\varepsilon, f(u))\}\,du + \int_{z-\varepsilon}^{z} c(u, f(u))\,du$$

$$- \int_{u^*}^{u^*+\varepsilon} c(u, f(z+u^*-u))\,du$$

$$> -\int_{u^*}^{z-\varepsilon} \varepsilon \left[\frac{\partial c}{\partial u}\right]_{(u, f(u))} \mathrm{d}u - \varepsilon^2 M$$

$$+ \varepsilon c(z, f(z)) - \varepsilon^2 M - \varepsilon c(u^*, f(z)) - \varepsilon^2 M,$$

using the bounds on the derivatives of c. Hence from the definition of r we have

$$\int_0^1 c(u, f(u))\,\mathrm{d}u - \int_0^1 c(u, f'(u))\,\mathrm{d}u$$

$$> -\varepsilon r(z-\varepsilon) + \varepsilon r(u^*) + \varepsilon c(z, v^*)$$

$$- \varepsilon c(u^*, v^*) - 3\varepsilon^2 M$$

$$> \varepsilon\{r(u^*) - r(z)$$

$$+ r(z) + s(v^*) - c(u^*, v^*)\} - 4\varepsilon^2 M$$

$$= \varepsilon\delta - 4\varepsilon^2 M,$$

as required. The proof goes through similarly when $z < u^*$. □

5.4.1 Convergence of the algorithm

The repeated application of the optimality check and improvement step, described above, constitute an algorithm for the continuous transportation problem. However, this algorithm might never reach an optimal solution. Indeed there are many examples where the only optimal solution is not a piecewise continuous assignment, and so could never be obtained by applying the algorithm. Nevertheless it is interesting to ask whether the objective-function value for the sequence of piecewise continuous assignments generated by the algorithm will converge to the optimal value. In this section we show that this is indeed the case provided that u^*, v^* and ε are chosen correctly, and provided that a modified definition of f' is made when f is discontinuous near z. This is an important result, suggesting as it does that similar properties may hold for other simplex extensions. Unfortunately, to obtain this result it is necessary to consider a number of technical details.

We make the following definitions. Set $f_0(u) = u$ and let f_n be the piecewise continuous assignment arrived at after n improvement steps starting at f_0. Set

$$c_n = \int_0^1 c(u, f_n(u))\,\mathrm{d}u,$$

$$\delta'_n = \sup_{u, v \in [0,1]} \{r_n(u) + s_n(v) - c(u, v)\},$$

where r_n and s_n are the reduced-cost functions for f_n. Finally let

$$\delta_n = \min\{\delta'_1, \delta'_2, \ldots, \delta'_n\}.$$

We can think of δ'_n as the maximum dual constraint dissatisfaction at stage n, or equivalently, the minimum value taken by the reduced-cost functional. Essentially, the content of the following result is that the right choice of pivot as we move from stage n to stage $n+1$ gives an improvement proportional to the square of this quantity.

Lemma 5.6 *We may choose u^*, v^* and ε at each iteration of the algorithm in such a way that*

(*a*) *the discontinuities of f_{n+1} are at least $\delta_n/(24M)$ apart, $n = 0,1,2,\ldots$, and*
(*b*) $c_n - c_{n+1} > \delta_n^2/(48\,M)$, $n = 0, 1, 2, \ldots$.

Proof. We prove (a) by induction; (b) follows from the choice of u^*, v^* and ε at each stage.

Suppose that the discontinuities of f_n occur at $u_1, u_2, \ldots, u_m$, with $0 = u_1 < u_2 < \cdots < u_m = 1$. From the definition of δ_n' there are sequences $\{u^{(i)}\}$ and $\{v^{(i)}\}$ with

$$r_n(u^{(i)}) + s_n(v^{(i)}) - c(u^{(i)}, v^{(i)}) \to \delta_n'$$

and, for some fixed p and q,

$$u_p < u^{(i)} < u_{p+1},\; u_q < f_n^{-1}(v^{(i)}) < u_{q+1}, \qquad i = 1,2,\ldots.$$

It follows from the bound on $\partial c/\partial u$ that, for every i,

$$2M\,|u^{(i)} - f_n^{-1}(v^{(i)})| > r_n(u^{(i)}) + s_n(v^{(i)}) - c(u^{(i)}, v^{(i)}), \tag{4}$$

for the quantity on the right-hand side of the inequality disappears if $u^{(i)}$ is replaced by $f_n^{-1}(v^{(i)})$.

Define $L = u_{q+1} - u_q$ and suppose first that $q > p$. For some N_u sufficiently large we can choose

$$u^* = u_p, \qquad \text{if } u_{p+1} - u_p < \delta_n/(12M),$$
$$u^* \in [u_p + \delta_n/(24M),\, u_{p+1} - \delta_n/(24M)], \qquad \text{otherwise},$$

so that $|u^* - u^{(i)}| \leqslant \delta_n/(12M)$ for every $i > N_u$. The choice of v^* and ε depends upon the size of L. By the inductive hypothesis $L \geqslant \delta_{n-1}/(24M) \geqslant \delta_n/(24M)$.

When $L \geqslant \delta_n/(6M)$, we choose v^* so that

$$u_q + \delta_n/(12M) \leqslant f_n^{-1}(v^*) \leqslant u_{q+1} - \delta_n/(12M)$$

and for some N_v, sufficiently large, $|v^* - v^{(i)}| \leqslant \delta_n/(12M)$ for every $i > N_v$. Let $N = \max(N_u, N_v)$. It follows from the bounds on the derivatives of c that, for every $i > N$,

$$|r_n(u^{(i)}) + s_n(v^{(i)}) - c(u^{(i)}, v^{(i)}) - r_n(u^*) - s_n(v^*) + c(u^*, v^*)|$$
$$< 2M\,|u^{(i)} - u^*| + 2M\,|v^{(i)} - v^*| \leqslant \delta_n/3.$$

Defining δ as in (3), it follows that

$$\delta = r_n(u^*) + s_n(v^*) - c(u^*, v^*) \geqslant \delta_n' - \delta_n/3 \geqslant 2\delta_n 3. \tag{5}$$

Let $\varepsilon = \delta_n/(24M)$, and construct f_{n+1} from f_n as described above. It is clear from

(4) that $\varepsilon < |z - u^*|$. Furthermore, from Theorem 5.5, it follows that

$$c_n - c_{n+1} > \delta_n \delta/(24M) - \delta_n^2/(144M) \geqslant (48M).$$

When $\delta_n/(12M) \leqslant L < \delta_n/(6M)$, we choose $v^* = f(u_q + L/2)$. Then, by a similar argument to that given above, (5) holds. Let $\varepsilon = L/2$, and construct f_{n+1} from f_n as above. From Theorem 5.5 it follows that

$$c_n - c_{n+1} > L\delta/2 - L^2 M \geqslant \delta_n^2/(48M).$$

When $\delta_n/(24M) \leqslant L < \delta_n(12M)$, we define f_{n+1} as follows:

$$\begin{aligned} f_{n+1}(u) &= f_n(u_{q+1} + u^* - u), && u \in (u^*, u^* + L), \\ &= f_n(u - L), && u \in (u^* + L, u_{q+1}), \\ &= f_n(u), && \text{otherwise.} \end{aligned}$$

Since $L < \delta_n/(12M)$, any choice of v^* with $f^{-1}(v^*) \in (u_q, u_{q+1})$ will satisfy (5). Using a similar proof to that of Theorem 5.5, it can be shown that

$$\begin{aligned} c_n - c_{n+1} &> L\delta - 4L^2 M \\ &\geqslant 2L\delta_n/3 - 4L^2 M \geqslant \delta_n^2/(48M). \end{aligned}$$

Suppose now that $q < p$. For N_u sufficiently large we can choose

$$\begin{aligned} &u^* = u_{p+1}, && \text{if } u_{p+1} - u_p < \delta_n/(12M), \\ &u^* \in [u_p + \delta_n/(24M), u_{p+1} - \delta_n/(24M)], && \text{otherwise} \end{aligned},$$

so that $|u^* - u^{(i)}| \leqslant \delta_n/(12M)$ for every $i > N_u$. The proof then proceeds as before, except for the case $\delta_n/(24\,M) \leqslant L < \delta_n/(12\,M)$; in this case we define f_{n+1} by

$$\begin{aligned} f_{n+1}(u) &= f_n(u_q + u^* - u), && u \in (u^* - L, u^*), \\ &= f_n(u + L), && u \in (u_q, u^* - L), \\ &= f_n(u), && \text{otherwise.} \end{aligned}$$

Finally we suppose that $q = p$. $L \geqslant \delta_n/(2M)$ from (4); we can thus choose $u^* \in [u_q + \delta_n/(24M), u_{q+1} - \delta_n/(24M)]$ so that for some large N_u, $|u^* - u^{(i)}| \leqslant \delta_n/(12M)$ for every $i > N_u$. Then v^* and ε are chosen as before, and the proof proceeds similarly.

In each case it is clear that the discontinuities of f_{n+1} are at least $\delta_n/(24M)$ apart. Since $f_0(u) = u$, this completes the proof. □

We can now prove the convergence result that we have been working towards.

Theorem 5.7 *The algorithm as applied in Lemma 5.6 converges to an optimal solution to CTP, in the sense that $c_n \to V(CTP)$.*

Proof. Since $V(\text{CTP}) \leqslant c_n$ for every n and c_n decreases, the sequence c_n must converge to $V(\text{CTP}) + k$ for some $k \geqslant 0$. Suppose $k > 0$; then by Lemma 5.5,

$\delta'_n > k$ for every n, and hence $\delta_n > k$, $n = 1,2,\ldots$. Thus

$$c_n - c_{n+1} > \delta_n^2/(48M) > k^2/(48M), \qquad n = 1,2,\ldots,$$

which gives a contradiction. Hence $k = 0$, as required. □

5.5 THE SEMI-INFINITE TRANSPORTATION PROBLEM

Our main interest in extensions of the transportation problem is in the fully infinite version. Nonetheless, the semi-infinite problem, in which the number of sources or the number of destinations is finite, is also of some interest. Examples of the problem arise in applications which involve meeting from a discrete set of sources a demand for a commodity from users distributed over an area. The problem has two interesting features: a defining property of its basic solutions, and the possibility of solving it by an interesting dual algorithm.

The problem, still called CTP, is formulated here with n sources and destinations distributed over the unit square

$$\text{CTP:}\quad \text{minimize} \quad \sum_{i=1}^{n} \int_0^1 \int_0^1 c_i(u_1, u_2)\rho_i(u_1, u_2)\,du_1\,du_2$$

$$\text{subject to} \quad \int_0^1 \int_0^1 \rho_i(u_1, u_2)\,du_1\,du_2 = a_i, \qquad i = 1, 2, \ldots, n,$$

$$\sum_{i=1}^{n} \rho_i(u_1, u_2) = 1, \qquad (u_1, u_2) \in [0, 1] \times [0, 1],$$

$$\rho_i(u_1, u_2) \geq 0, \qquad (u_1, u_2) \in [0, 1] \times [0, 1].$$

We assume that each function c_i is continuous. To conform strictly with our earlier formulation, the variable space would have to be the space of regular Borel measures on $\{1,2,\ldots,n\} \times ([0,1] \times [0,1])$. For simplicity, we have posed the problem in the space of absolutely continuous measures; the constraints then ensure that any feasible solution will be in L_∞. We choose $R^n \times C([0,1] \times [0,1])$ for the dual variable space, and $\{C([0,1] \times [0,1])\}^n$ for the dual constraint space. In this setting, the dual problem, again called CTP*, is

$$\text{CTP*:}\quad \text{maximize} \quad \sum_{i=1}^{n} a_i s_i + \int_0^1 \int_0^1 r(u_1, u_2)\,du_1\,du_2$$

$$\text{subject to} \quad r(u_1, u_2) + s_i \leqslant c_i(u_1, u_2),$$

$$(u_1, u_2) \in [0, 1] \times [0, 1], \quad i = 1, 2, \ldots, n.$$

The place of the compact sets U and V is here taken by $\{1,2,\ldots,n\}$ and $[0,1] \times [0, 1]$ respectively. Theorem 5.2 ensures that strong duality holds for this version of CTP and CTP*. In the next two theorems we characterize the basic solutions for CTP and CTP*. We begin by showing that a basic solution for CTP is necessarily an assignment; that is, for each (u_1, u_2), all but one of the functions ρ_i takes the value zero.

Theorem 5.8 *A feasible solution ρ for* CTP *is basic if and only if for almost all $(u_1, u_2) \in [0,1] \times [0,1]$, exactly one of the functions ρ_i, $i = 1,2,\ldots,n$, is non-zero.*

Proof. Suppose Γ is a set of non-zero measure on which ρ_i and ρ_j are non-zero. Then there exists a subset, still called Γ, and $k > 0$ such that Γ has non-zero measure and both ρ_i and ρ_j exceed k almost everywhere in Γ. Now divide Γ into two sets Γ_1, Γ_2 of equal measure, and define a measure μ on $\{1,2,\ldots,n\} \times ([0,1] \times [0,1])$, by setting $\mu_k = 0$ unless $k = i$ or $k = j$, and

$$\begin{aligned} \mu_i(u_1, u_2) &= k, && (u_1, u_2) \in \Gamma_1, \\ &= -k, && (u_1, u_2) \leqslant \in \Gamma_2, \\ &= 0, && \text{otherwise}, \end{aligned}$$

$$\mu_j = -\mu_i.$$

Then it is clear that $\rho + \mu$ and $\rho - \mu$ are both feasible, and hence ρ cannot be basic. The reverse implication is obvious. □

As we have already remarked, the constraints of CTP* are such that, given any feasible solution (r, s), we can add an arbitrary constant function to r and subtract this constant from s to produce another feasible solution. Thus, in general, there are no basic feasible solutions for the constraints as they stand. We can remove this degeneracy by adding the constraint $s_1 = 0$ without materially changing the problem. If we do this, we can characterize the basic solutions of CTP* in a way which will allow us to formulate a simple dual algorithm for CTP. From now on, we assume that the constraints of CTP* are augmented in this way.

Theorem 5.9 *If (r, s) is a feasible solution for* CTP*, *then (r, s) is basic if and only if*

$$r(u_1, u_2) = \min_i \{c_i(u_1, u_2) - s_i\}, \qquad \text{for all } 0 \leqslant u_1, u_2 \leqslant 1,$$

$$s_i = \min_{0 \leqslant u_1, u_2 \leqslant 1} \{c_i(u_1, u_2) - r(u_1, u_2)\}, \qquad i = 1,2,\ldots,n.$$

Proof. Suppose first that there is some i such that $s_i < c_i(u_1, u_2) - r(u_1, u_2)$ for every $u_1, u_2 \in [0,1]$. Then if $i = 1$, there is some $\varepsilon > 0$ such that $(r + r_\varepsilon, s + s_\varepsilon)$ and $(r - r_\varepsilon, s - s_\varepsilon)$ are both feasible, where $r_\varepsilon(u_1, u_2) = \varepsilon$, $0 \leqslant u_1, u_2 \leqslant 1$ and $s_\varepsilon = (0, -\varepsilon, -\varepsilon, \ldots, -\varepsilon)$. If $i > 1$, then we may simply add or subtract ε to s_i, leaving everything else unchanged, to yield two new feasible solutions. In either case, this shows that (r, s) is not basic.

Suppose, on the other hand, that there is $v_1, v_2 \in [0,1]$ with $r(v_1, v_2) < c_i(v_1, v_2) - s_i$, $i = 1,2,\ldots,n$. Define a function r_ε on $[0,1] \times [0,1]$ by

$$r_\varepsilon(u_1, u_2) = \max\{0, \varepsilon - (u_1 - v_1)^2 - (u_2 - v_2)^2\}.$$

From the continuity of c and r it is straightforward to deduce that $(r + r_\varepsilon, s)$ and

$(r - r_\varepsilon, s)$ are both feasible for $\varepsilon > 0$ chosen small enough. So (r, s) is not basic.

Suppose now that the conditions of the theorem hold, but that (r, s) is not basic. Then there exist r', s' with $s'_1 = 0$ and such that $(r + r', s + s')$ and $(r - r', s - s')$ are both feasible. This implies that, for each i, $r'(u_1, u_2) = -s'_i$ throughout the set of points where $c_i(u_1, u_2) = r(u_1, u_2) + s_i$. These sets are non-empty and together cover the whole unit square. Hence r' takes each of the values $\{-s'_i, i = 1, 2, \ldots, n\}$, and no other values. Since r' is continuous, all these values must be equal, and hence zero since $s'_1 = 0$. Hence both r' and s' are zero. □

This theorem is interesting because it completely defines the basic solutions and because, surprisingly, the analogous result is not true in either the fully infinite case or the finite case. An example of an infinite transportation problem where the theorem does not hold is given by $c(u, v) = u + |u - v|, 0 \leqslant u \leqslant 1, 0 \leqslant v \leqslant 1$. Then the solution

$$r(u) = u, \qquad 0 \leqslant u \leqslant 1,$$

$$s(v) = 0, \qquad 0 \leqslant v \leqslant 1,$$

is not basic, for if we define $r'(u) = u, 0 \leqslant u \leqslant 1$, and $s'(v) = -v, 0 \leqslant v \leqslant 1$, then both $(r + r', s + s')$ and $(r - r', s - s')$ are feasible. It is easily verified, however, that

$$r(u) = \min_v \{c(u, v) - s(v)\},$$

$$s(v) = \min_u \{c(u, v) - r(u)\}.$$

Theorem 5.9 allows us to write r as a function of s for any basic feasible solution (r, s) of CTP*. This can be used as the starting point for a dual algorithm for CTP. Given r and s, we can generate a function ρ on the unit square which is complementary to (r, s), by putting $\rho_i(u_1, u_2) = 1$ wherever $c_i(u_1, u_2) - s_i - r(u_1, u_2)$ is zero, and $\rho_i = 0$ elsewhere. We assume that these regions do not overlap, so that exactly one component of ρ is non-zero at any point (if they do overlap, we can assign the value 1 to the lowest numbered component of ρ amongst those eligible in any region, and zero to the others). The function ρ will not necessarily be feasible for CTP. If it is, then it is optimal, by weak duality and complementarity. If it is not, then the constraint violations will indicate appropriate changes in s.

It is easier to implement this algorithm in the original setting of CTP*, where s_1 is unrestricted. The basic solutions of the augmented constraints correspond to equivalence classes of solutions; (r, s) and (r', s') belong to the same equivalence class if $r + s$ and $r' + s'$ are the same function on $[0,1] \times [0,1] \times \{1, 2, \ldots, n\}$. Thus $(r - r')$ and $(s - s')$ are constant functions on $([0,1] \times [0,1])$ and $\{1, 2, \ldots, n\}$ respectively whenever (r, s) and (r', s') belong to the same equivalence class.

In outline, the steps of the algorithm are as follows.

(1) Choose s arbitrarily.
(2) Compute $r(u_1, u_2) = \min \{c_i(u_1, u_2) - s_i\}$, for all $(u_1, u_2) \in [0, 1] \times [0, 1]$.

(3) Compute ρ complementary to (r, s) as described above. Let M_i be the area of supp (ρ_i), $i = 1,2, \ldots, n$.
(4) If $M_i = a_i$, $i = 1,2, \ldots, n$ then stop: the current ρ is optimal for the primal and (r, s) is optimal for the dual. Otherwise choose k such that $M_i - a_i \leqslant M_k - a_k$ for all i. Increase s_k leaving the rest of s fixed and go to step 2.

The difficulty in defining this algorithm more precisely is in knowing how much to increase s_k in step 4. One possibility is to recalculate M_k as s_k changes and stop when $M_k = a_k$.

Actually, this algorithm is just equivalent to a particular implementation of an ascent algorithm for the problem one obtains by using the relation

$$r(u_1, u_2) = \min_i \{c(u_1, u_2)s_i\}$$

to eliminate r from CTP*, which then becomes an unconstrained maximization over n variables. It is left for the reader to confirm that the vector whose kth component is $M_k - a_k$ is the gradient of the objective function for this reduced problem. The algorithm specified above is equivalent to solving this unconstrained problem by one-variable-at-a-time ascent.

5.6 NOTES

The mass-transfer problem was first studied by the French mathematician, Gaspard Monge (1781). In his analysis of this problem, Monge reformulated it in terms of descriptive geometry. His results, and those of his successors Charles Dupin (1818) and Albert de St. Germaine (1886), were primarily relevant to differential and descriptive geometry and left many open questions regarding the existence and nature of solutions to the problem. In 1885 Paul Appell won half the Bordin Prize, offered by the Academy of Sciences of Paris for a solution to the problem. His subsequent papers of 1887 and 1928 critically examined and extended the work of Monge. He looked in detail at the implications of the fact that optimal routes may cross each other only at their endpoints. This result, which was originally established by Monge, is a consequence of the metric properties of the cost function c.

The mass-transfer problem was next examined by Kantorovich (1942), who posed the problem as a mathematical program over a space of functions on the Borel sets of a compact metric space. Although Kantorovich did not explicitly formulate a dual problem, he stated and proved what would now be described as a strong duality result, for the case where c is any continuous non-negative function. This proof however is obscure and is now suspected to be invalid (see Kortanek and Yamasaki (1980)). In a second paper, Kantorovich and Rubinstein (1958) restated the strong duality result for the case where c is a metric. The significance of Kantorovich's paper lies in the formulation of the problem as a mathematical program over a function space. This abstract approach, distinct from the descriptive work of the French geometers, has been adopted and continued in the modern literature; see Levin (1975), Levin and Milyutin (1978) and Rachev (1985).

The semi-infinite transportation problem has been considered by a number of authors (Corley and Roberts (1972), Lowe and Hurter (1976)). The duality theory of this problem has been discussed by Todd (1978) and by Kortanek and Yamasaki (1982). A number of the results we have given on extreme points of transportation problems were first established by Lewis (1985). He also proves that Theorem 5.9 can be extended to the fully infinite case under an assumption on the differentiability of the functions involved. Much of the material of Sections 5.3 and 5.4 appears in Philpott (1982) and Anderson and Philpott (1984).

Problems

5.1 Show that the positive cone in $M_r(U \times V)$ is closed in the $\sigma(M_r(U \times V), C(U \times V))$ topology.

5.2 Consider the problem CTP with $U = V = [0, 1]$ and μ_1 and μ_2 some absolutely continuous measures on U and V, not equal to Lebesgue measure. Show how this can be transformed into a problem in the form CTP, with both μ_1 and μ_2 being Lebesgue measure.

5.3 Prove that if r and s are the reduced cost functions for an optimal solution ρ, then s is continuous.

5.4 Consider the problem with cost function $c(u, v) = u^2v - uv^2$ and initial solution given by $f(u) = 1 - u$. Calculate the reduced cost functions and hence make one improvement step. Check that your solution has a smaller objective function value.

5.5 Consider the problem of finding the joint distribution of random variables X and Y, each defined on R, which maximizes $E(XY)$, given fixed marginal distributions for both X and Y. Express this as a continuous transportation problem and find its dual.

5.6 Prove that the solution suggested for the example shown in Figure 5.1 is indeed optimal.

6

Maximal Flow in a Dynamic Network

6.1 INTRODUCTION

In this chapter we consider an infinite-dimensional extension of the problem of finding the maximal flow in a network. In its classical form this is the problem of choosing flows of some commodity in a directed network in such a way as to maximize the total flow between one particular node (called the source) and another (called the sink) subject to a fixed set of constraints, in the form of upper bounds on the flows in the arcs. This problem can be formulated as the following linear program with n^2 variables.

$$\begin{aligned}
\text{NP:}\quad & \text{maximize} && \sum_{k=1}^{n} [x_{kn} - x_{nk}] && \\
& \text{subject to} && \sum_{k=1}^{n} [x_{jk} - x_{kj}] = 0, && j = 2,3,\ldots,n-1, \quad (1) \\
& && 0 \leqslant x_{jk} \leqslant b_{jk}, && j,k = 1,2,\ldots,n. \quad (2)
\end{aligned}$$

Here node 1 is the source, node n is the sink, and x_{jk} is the flow in the arc (having capacity b_{jk}) from node j to node k.

We consider the maximal *dynamic* flow problem in a continuous-time framework. The problem is posed over a time interval, and the nodes of the network have storage capacities. Thus a net flow into some node during one period may be followed by a net flow out of that node during a later period. The flows and the storage vary as functions of time, bounded above by arc capacities and storage capacities respectively, which are also functions of time.

This problem is an infinite-dimensional linear program. However, the treatment we give is somewhat different in style to the previous chapters. The reason is that our approach is modelled on that of Ford and Fulkerson, whose elegant and famous maximum-flow minimum-cut theorem is the basis of a labelling algorithm which is only tangentially related to the simplex algorithm. Our discussion is motivated by considering a discretized version of the maximal dynamic flow problem. This leads to the development of a continuous-time analogue of the

Ford–Fulkerson theorem. We also discuss duality theory for the time-continuous problem.

6.2 THE TIME-CONTINUOUS NETWORK-FLOW PROBLEM

We define a network Ω by a set S of nodes, numbered 1 through n, where node 1 is the source and node n is the sink, and a set A of directed arcs. We assume that all the possible arcs are present, so that $A = S \times S$, and we write (j, k) for the arc going from node j to node k. With each arc (j, k) we associate a non-negative bounded measurable function b_{jk} on $[0, T]$ which gives its capacity at each time, and with each node j we associate a continuous function a_j on $[0, T]$ which gives its storage capacity.

We wish to find a set of flows in each arc, that is non-negative functions x_{jk} on $[0, T]$, so as to maximize the volume, v, of material flowing through the network into node n over $[0, T]$. Clearly

$$v = \int_0^T \sum_{k=1}^n [x_{kn}(t) - x_{nk}(t)]\,\mathrm{d}t.$$

Now x_{jk}, the flow in arc (j, k), is subject to conservation constraints at each node and is bounded by the capacity b_{jk}. Thus the time-continuous network-flow problem (TCNP) can be posed as:

$$\text{TCNP:}\quad \text{maximize} \quad \int_0^T \sum_{k=1}^n [x_{kn}(t) - x_{nk}(t)]\,\mathrm{d}t$$

$$\text{subject to} \quad \int_0^t \sum_{k=1}^n [x_{kj}(\tau) - x_{jk}(\tau)]\,\mathrm{d}\tau \leqslant a_j(t),$$

$$j = 2,3, \ldots, n-1, \tag{3}$$

$$\int_0^t \sum_{k=1}^n [x_{jk}(\tau) - x_{kj}(\tau)]\,\mathrm{d}\tau \leqslant 0,$$

$$j = 2,3, \ldots, n-1, \tag{4}$$

$$x_{jk}(t) \leqslant b_{jk}(t), \tag{5}$$

$$x_{jk}(t) \geqslant 0, \quad j, k = 1,2, \ldots, n \quad t \in [0, T]. \tag{6}$$

Here the flows x_{jk} are members of $L_\infty[0, T]$. We say a set $\{x_{jk}\}$ of flows is a *feasible flow* if constraints (3)–(6) are satisfied. The *value* of the flow is the volume v with which it corresponds.

We take the primal variables to lie in $L_\infty\,[0, T]$ and the dual constraint space, Y, to be an L_1 space. Then, because of the constraints (5), the feasible region is bounded in the primal variable space and so we can immediately apply Theorem 3.20 (strictly speaking its inequality-constrained counterpart) to guarantee the existence of an optimal solution to TCNP.

To motivate our treatment of this problem, we begin by considering a discrete version of the time-continuous network-flow problem for which a cut can be defined by an obvious extension of the classical definition. By observing this cut for successively finer discretizations, it is possible to deduce the nature of the cut for the continuous problem.

A discrete version, DNP, of the continuous problem can be formulated over N discrete time epochs, $t = 1,2, \ldots, N$, as follows:

$$\text{DNP:} \quad \text{maximize} \quad \sum_{t=1}^{N} \sum_{k=1}^{n} [x_{kn}(t) - x_{nk}(t)]$$

$$\text{subject to} \quad \sum_{i=1}^{t} \sum_{k=1}^{n} [x_{kj}(i) - x_{jk}(i)] \leqslant a_j(t),$$

$$\sum_{i=1}^{t} \sum_{k=1}^{n} [x_{jk}(i) - x_{kj}(i)] \leqslant 0, \quad j = 2,3, \ldots, n-1,$$

$$x_{jk}(t) \leqslant b_{jk}(t),$$

$$x_{jk}(t) \geqslant 0, \quad j, k = 1,2, \ldots, n, \quad t = 1,2, \ldots, N.$$

It is convenient to modify the constraints of DNP by introducing variables $y_j(t)$, which measure the amount of storage in each node. We define $y_j(0) = 0$ for $j = 2,3, \ldots, n-1$, to obtain the equivalent problem:

$$\text{maximize} \quad \sum_{t=1}^{N} \sum_{k=1}^{n} [x_{kn}(t) - x_{nk}(t)]$$

$$\text{subject to} \quad \sum_{k=1}^{n} [x_{jk}(t) - x_{kj}(t)] + y_j(t) - y_j(t-1) = 0,$$

$$j = 2,3, \ldots, n-1,$$

$$y_j(t) \leqslant a_j(t), \quad j = 2,3, \ldots, n-1,$$

$$y_j(t) \geqslant 0, \quad j = 2,3, \ldots, n-1,$$

$$x_{jk}(t) \leqslant b_{jk}(t),$$

$$x_{jk}(t) \geqslant 0, \quad j, k = 1,2, \ldots, n, \quad t = 1,2, \ldots, N.$$

Expanding the network in a similar fashion to that proposed by Ford and Fulkerson in their discussion of maximal dynamic flow, we obtain the *time-augmented* extension Ω_T of Ω. This is defined by a set of nodes

$$S_T = \{(i, t): \; i \in S, t = 1,2, \ldots, N\} \cup \{p, q\}$$

and a set of arcs A_T joining the nodes of S_T defined by the rules:

(a) There is an arc of capacity $b_{jk}(t)$ joining node $(j, t) \in S_T$ and $(k, t) \in S_T$ for $t = 1, 2, \ldots, N$.

(b) There is an arc of infinite capacity joining node p to $(1, t) \in S_T$ for $t = 1,2, \ldots, N$.
(c) There is an arc of infinite capacity joining node $(n, t) \in S_T$ to node q for $t = 1,2, \ldots, N$.
(d) There is an arc of capacity $a_j(t)$ joining node $(j, t) \in S_T$ to node $(j, t+1) \in S_T$ for $t = 1,2, \ldots, N-1$ and $j = 2,3, \ldots, n-1$.
(e) There is an arc of capacity $a_j(N)$ joining node $(j, N) \in S_T$ to node $p \in S_T$ for $j = 2,3, \ldots, n-1$.

An example of a network Ω_T produced by this procedure is shown in Figure 6.1.

It can be seen quite readily that DNP is the problem of maximizing the flow through the network Ω_T into node q subject to the capacity constraints on the arcs in A_T. The arcs defined by (e) ensure that flow into the sink is maximized at the

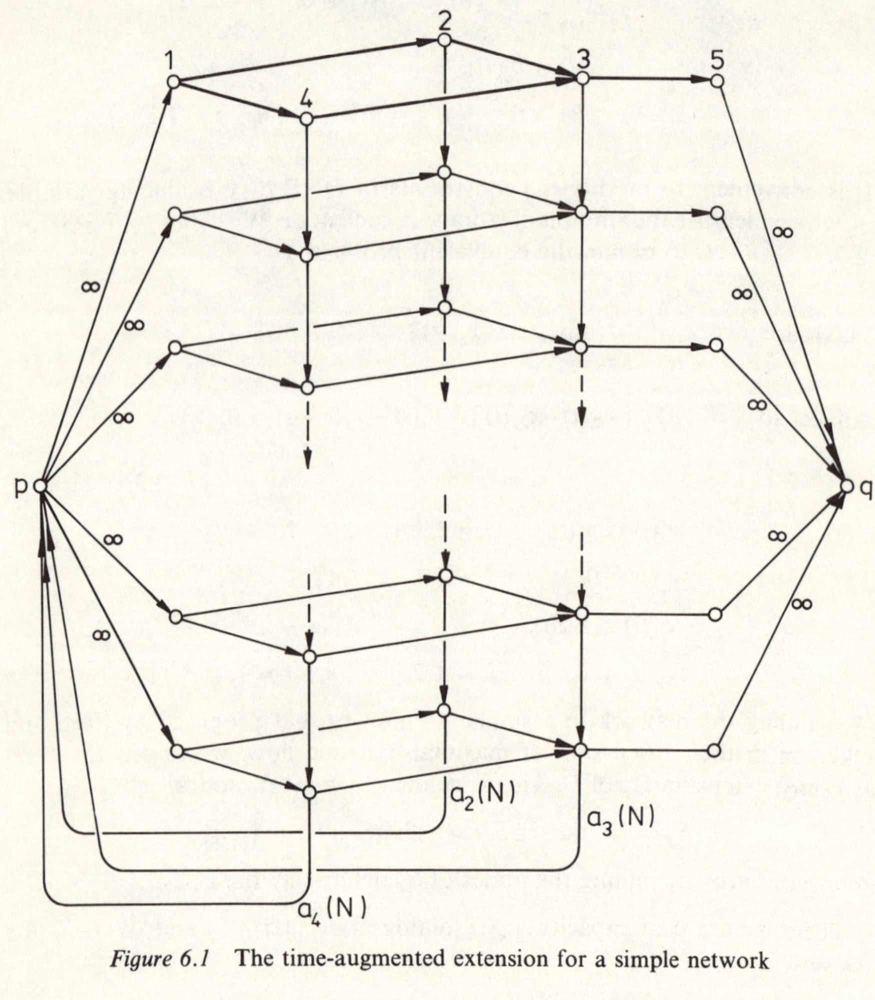

Figure 6.1 The time-augmented extension for a simple network

expense of flows which produce residual storage at the final time. This problem is static in nature and can be solved using the classical labelling algorithm to give an optimal flow. There is a corresponding minimum cut defined by a partition of S_T into a set X defined by the labelled nodes of Ω_T, and its complement $\tilde{X}$ (throughout this chapter we use the notation $\tilde{B}$ to represent the complement of any set B). This cut 'crosses' only those arcs having full capacity flow towards the sink and zero reverse flow; we can deduce that it crosses a node j (or strictly the arc of A_T joining node (j, t) to $(j, t+1)$) only when the storage in j at time t is full (when node j passes from the source side of the cut to the sink side) or empty (when node j passes from the sink side of the cut to the source side).

The value V of this cut is given by the sum over j in X and k in $\tilde{X}$ of the capacities of the arcs in A_T joining the nodes j and k. Note that in general j and k are ordered pairs of the form (j, t), for some $t = 1, 2, \ldots, N$, and that if either $(1, t) \in \tilde{X}$ or $(n, t) \in X$ for any t, then $V = \infty$. Otherwise define

$$S_1 = \{(j, k, t): (j, t) \in X, \quad (k, t) \in \tilde{X}\},$$

$$S_2 = \{(j, t): (j, t) \in X, \quad (j, t+1) \in \tilde{X}\}.$$

Then

$$V = \sum_{(j, k, t) \in S_1} b_{jk}(t) + \sum_{(j, t) \in S_2} a_j(t).$$

This expression gives an indication of how we should define a cut in the continuous version of the problem. By considering successively finer discretizations, we can see that for the continuous problem the value of a cut comprises an integral of arc capacities over some specified subsets of $[0, T]$, which correspond to S_1, and a sum of storage capacities over a set which corresponds to S_2. To make this apparently straightforward construction precise requires some technical definitions which will now be discussed in detail.

6.3 GENERALIZED CUTS

A *generalized cut* C in Ω separating nodes 1 and n is a set-valued function of time on $[0, T]$, with $C(t)$ a subset of S such that for every $t \in [0, T]$, $1 \in C(t)$ and $n \in \tilde{C}(t)$, and for each j in S, the set

$$\Gamma_j = \{t: j \in C(t)\} \quad (0, T)$$

is open. Since Γ_j is open we can express it as a countable union of disjoint open intervals, unique up to order (see, for example, Apostol (1957), Theorem 3.9). Hence

$$\Gamma_j = \bigcup_{i=1}^{\infty} (\alpha_{ij}, \beta_{ij}).$$

For each node j we let P_j and Q_j be the sets of left- and right-hand end points, respectively, of these open intervals. Thus

$$P_j = \{\alpha_{ij}: i = 1,2, \ldots\},$$

$$Q_j = \{\beta_{ij}: i = 1,2, \ldots\}.$$

Moreover we let $R_j = Q_j - \{T\}$ if $T \in Q_j$, and $R_j = Q_j$ otherwise.

Now we are ready to define the *value* V of the cut C as

$$V = \sum_{j=1}^{n} \sum_{k=1}^{n} \int_{\Gamma_j \cap \tilde{\Gamma}_k} b_{jk}(\tau)\,\mathrm{d}\tau + \sum_{j=2}^{n-1} \sum_{t \in R_j} a_j(t),$$

if $\sum_{j=2}^{n-1} \sum_{t \in R_j} a_j(t)$ exists, and infinity otherwise.

Theorem 6.1 *Suppose that v is the value of some flow which is feasible for TCNP and V is the value of some generalized cut C in Ω. Then $v \leqslant V$.*

Proof. Let $y_j(t)$ be the storage in node j at time t. Thus

$$y_j(t) = \int_0^t \sum_{k=1}^{n} [x_{kj}(\tau) - x_{jk}(\tau)]\,\mathrm{d}\tau, \qquad j = 2, 3, \ldots, n-1.$$

Since these equations hold for every $t \in [0, T]$, by restricting the region of integration to Γ_j for each j, we can deduce

$$\int_{\Gamma_j} \sum_{k=1}^{n} [x_{kj}(\tau) - x_{jk}(\tau)]\,\mathrm{d}\tau = \sum_{i=1}^{\infty} [y_j(\beta_{ij}) - y_j(\alpha_{ij})], \tag{7}$$

$$j = 2, 3, \ldots, n-1,$$

when the sum on the right exists. In any case, since $y_j(t) \geqslant 0$ for every $t \in [0, T]$,

$$\int_{\Gamma_j} \sum_{k=1}^{n} [x_{kj}(\tau) - x_{jk}(\tau)]\,\mathrm{d}\tau \leqslant \sum_{i=1}^{\infty} y_j(\beta_{ij}), \tag{8}$$

$$j = 2, 3, \ldots, n-1.$$

Since $\Gamma_1 = (0, T)$, we have, from the definition of v and y_j,

$$v = \int_{\Gamma_1} \sum_{k=1}^{n} [x_{1k}(\tau) - x_{k1}(\tau)]\,\mathrm{d}\tau - \sum_{j=2}^{n-1} y_j(T). \tag{9}$$

Thus from (8), (9), and $\Gamma_n = \varnothing$,

$$v \leqslant \sum_{j=1}^{n} \int_{\Gamma_j} \sum_{k=1}^{n} [x_{jk}(\tau) - x_{kj}(\tau)]\,\mathrm{d}\tau + \sum_{j=2}^{n-1} \left[\sum_{i=1}^{\infty} y_j(\beta_{ij}) - y_j(T) \right].$$

From the definition of Γ_j, it can be seen that

$$\begin{aligned} &\sum_{j=1}^{n} \int_{\Gamma_j} \sum_{k=1}^{n} [x_{jk}(\tau) - x_{kj}(\tau)]\,\mathrm{d}\tau \\ &\quad = \sum_{j=1}^{n} \sum_{k=1}^{n} \int_{\Gamma_j} [x_{jk}(\tau) - x_{kj}(\tau)]\,\mathrm{d}\tau \\ &\quad = \sum_{j=1}^{n} \sum_{k=1}^{n} \int_{\Gamma_j \cap \tilde{\Gamma}_k} [x_{jk}(\tau) - x_{kj}(\tau)]\,\mathrm{d}\tau \end{aligned}$$

$$\leqslant \sum_{j=1}^{n} \sum_{k=1}^{n} \int_{\Gamma_j \cap \bar{\Gamma}_k} b_{jk}(\tau)\,\mathrm{d}\tau.$$

Also, since the constraints (3) of TCNP ensure that $y_j(t) \leqslant a_j(t)$, it follows that

$$\sum_{j=2}^{n-1}\left[\sum_{i=1}^{\infty} y_j(\beta_{ij}) - y_j(T)\right] \leqslant \sum_{j=2}^{n-1}\sum_{t\in R_j} y_j(t)$$

$$\leqslant \sum_{j=2}^{n-1}\sum_{t\in R_j} a_j(t).$$

This establishes that $v \leqslant V$. □

Hence the value of any generalized cut gives an upper bound on the value of any feasible flow. This is reminiscent of a linear programming weak duality result and we shall see in Section 6.5 the way in which these results are connected with linear programming duality.

In the next section we consider the possibility of constructing a feasible flow and a generalized cut having equal values, thus solving TCNP and proving a maximum-flow minimum-cut theorem. However, we should not expect this theorem to hold with full generality in this setting, and we now give an example to illustrate this.

Consider a three-node network over the time interval [0, 1], where node 1 is the source, node 3 is the sink and there are arcs (1, 2) and (2, 3). Suppose $a_2(t) = 0$ and $b_{23}(t) = 1$, $t \in [0, 1]$. Let $\{q_1, q_2, \ldots\}$ be an enumeration of the rationals and let the measurable set Q be defined by

$$Q = \left[\bigcup_{n=1}^{\infty} (q_n - 2^{-n-2},\, q_n + 2^{-n-2})\right] \cap [0,1].$$

Let $R = [0, 1] - Q$. We define $b_{12}(t) = 0$, $t \in Q$, and $b_{12}(t) = 2$, $t \in R$. Then $m(R)$, the measure of R, is at least $\frac{1}{2}$ and it is not hard to see that $m(R)$ is the value of the maximum flow. However, the optimal cut we would wish to define has $\Gamma_2 = R$, which is not open (for it contains none of the rationals). There are a number of ways in which we might try to overcome this problem, but it turns out to be extremely hard to produce an entirely satisfactory theory. In this chapter we prove a form of the maximum-flow minimum-cut theorem with a piecewise continuity restriction and then establish, using a different method, a maximum-flow infimum-cut theorem.

6.4 REACHABLE NODES

Suppose the flows $\{x_{jk}\}$ in the network Ω are feasible for TCNP. The *reachable* nodes of Ω are defined by the finite application of the following recursive rules:

(a) Node 1 is reachable for every time $t \in [0, T]$.

(b) If node j is reachable at time t and there is some $\delta > 0$ such that

$$\operatorname*{ess\,inf}_{t-\delta<\tau<t+\delta} \{b_{jk}(\tau)-x_{jk}(\tau)\} > 0,$$

then k is reachable at time t.

(c) If node j is reachable at time t and there is some $\delta > 0$ such that

$$\operatorname*{ess\,inf}_{t-\delta<\tau<t+\delta} x_{kj}(\tau) > 0,$$

then k is reachable at time t.

(d) If node j is reachable at time t_1 and for every $t \in [t_1, t_2)$, $y_j(t) < a_j(t)$, then j is reachable at all times $t \in (t_1, t_2)$.

(e) If node j is reachable at time t_1 and for every $t \in (t_2, t_1]$, $y_j(t) > 0$, then j is reachable at all times $t \in (t_2, t_1)$.

(f) If $y_j(T) > 0$, then node j is reachable from node 1 at time T.

This definition warrants some discussion. In general, the reachable nodes of the network are intended to be those having chains of nodes and arcs connecting them to the source, through which a feasible increase in flow can be made. This is not the case when rule (f) is invoked; here the increase in flow comes from residual storage in the network at time T which can then flow into the sink through arcs defined by the other rules. In both cases the improvement in flow must be made on some set of non-zero measure; for this reason the definition is such that nodes reachable at time t are reachable everywhere in some interval containing t. These properties are made precise by Lemma 6.2 below.

In the classical network-flow problem, the optimal cut is specified by the set of nodes which are labelled when the labelling algorithm terminates at the optimal flow. The same procedure can be followed for the time-continuous problem, where the reachable nodes assume the role of the labelled nodes. Termination of the continuous-labelling algorithm occurs when node n is not reachable at any time; this then determines a generalized cut which has an optimal value. This relationship is made precise by Lemma 6.3. From this and the previous results we then deduce a continuous version of the Ford–Fulkerson theorem.

Lemma 6.2 *Suppose there is a feasible flow with the sink (node n) reachable at some time $t \in (0, T)$. Then it is possible to construct a feasible flow with a value strictly greater than that of the existing flow.*

Proof. Suppose that, for the feasible flow $\{x_{jk}\}$, node n is reachable at some time by the successive applications of rules (b)–(f). Then there is some finite list

$$R = \{t_0, j_0, t_1, j_1, \ldots, t_q, j_q\},$$

where $j_0 = 1$ and $j_q = n$, such that j_i is reachable at time t_i by virtue of the fact that j_{i-1} is reachable at t_{i-1} and a single application of one of the rules (b)–(f). Without loss of generality we suppose that all circuits and redundant nodes have been deleted from R, so that

(A) if $j_{i-1} = j_i$, then $j_{i-2} \neq j_i$, $j_{i+1} \neq j_i$, and $t_{i-1} \neq t_i$;
(B) if $j_i = j_{i+1} = j_k = j_{k+1}$, where $k \neq i$, then the intervals having endpoints t_i, t_{i+1} and t_k, t_{k+1} are disjoint.

We consider first the case when $t_1 \neq T$ and hence rule (f) has not been used in the generation of R. Consider the successive applications of the other rules to consecutive nodes j_{i-1}, j_i, in R. Let $j = j_{i-1}$ and $k = j_i$. For an application of (b) or (c) at time t_i, say, we can find a $\delta > 0$ and a $z > 0$ such that either (for (b))

$$x_{jk}(s) + z \leqslant b_{jk}(s), \qquad \text{for almost every } s \in (t_i - \delta,\ t_i + \delta),$$

or (for (c))

$$x_{kj}(s) \geqslant z \qquad \text{for almost every } s \in (t_i - \delta,\ t_i + \delta).$$

Since a_j and y_j are continuous on $[0, T]$, for each application of (d) or (e) (when $j_{i-1} = j_i = j$) we can find an $\varepsilon > 0$ and a $w > 0$ such that either (for (d))

$$y_j(s) + w \leqslant a_j(s), \qquad s \in (t_{i-1} - \varepsilon, t_i + \varepsilon),$$

or (for (e))

$$y_j(s) \geqslant w, \qquad s \in (t_i - \varepsilon,\ t_{i-1} + \varepsilon).$$

Thus if n is reachable, there is a sequence of values of δ and z (say $\delta^{(1)}$, $z^{(1)}$, $\delta^{(2)}$, $z^{(2)}, \ldots, \delta^{(m)}, z^{(m)}$), and a sequence of values of ε and w (say $\varepsilon^{(1)}, w^{(1)}, \varepsilon^{(2)}, w^{(2)}, \ldots,$ $\varepsilon^{(p)}, w^{(p)}$), defined from the finite application of rules (b)–(e) to nodes in R. Choose $\delta^* > 0$ and $z^* > 0$ such that

$$\delta^* < \delta^{(i)},\ z^* < z^{(i)}, \qquad i = 1,2, \ldots, m,$$

$$\delta^* < \varepsilon^{(i)},\ 2\delta^* z^* = w^* < w^{(i)}, \qquad i = 1,2, \ldots, p.$$

Construct a new flow $\{x'_{jk}\}$ from $\{x_{jk}\}$ as follows. If $j = j_{i-1}$, $t = t_i$, $k = j_i$ are members of R, where $j \neq k$, then if rule (b) has been applied at t on arc (j,k),

$$x'_{jk}(s) = x_{jk}(s) + z^*, \qquad s \in (t - \delta^*,\ t + \delta^*)$$

and if rule (c) has been applied at t on arc (j,k),

$$x'_{kj}(s) = x_{kj}(s) - z^*, \qquad s \in (t - \delta^*,\ t + \delta^*).$$

Otherwise x'_{jk} is equal to x_{jk}. Clearly, the value of the new flow is strictly greater than that of the original flow by an amount w^*. Also, $x'_{jk}(s) \leqslant b_{jk}(s)$ and $x'_{kj}(s) \geqslant 0$ for almost every s where $\{x'_{jk}\}$ differs from $\{x_{jk}\}$. Thus the arc constraints (5) and (6) of TCNP are satisfied almost everywhere by the new flow. It is clear that we can change each x_{jk} on a set of measure zero to satisfy (5) and (6) everywhere.

Now let j be any node of Ω. If j is not a member of R, then $y'_j = y_j$ and is such that the constraints (3) and (4) of TCNP are satisfied trivially. Otherwise, it follows that $y'_j(t - \delta^*) = y_j(t - \delta^*)$, where t is the smallest t_i in R having $j_i = j$. We proceed to show that y'_i differs from y_j only on intervals having endpoints

$\min\{t_{i-1}, t_i\}-\delta^*$, $\max\{t_{i-1}, t_i\}+\delta^*$, where i is such that $j_i = j_{i-1} = j$. Furthermore, we show that y'_j is such that the storage constraints (3) and (4) of TCNP are satisfied.

Clearly if $j_{i-1} \neq j_i = j \neq j_{i+1}$ are members of R, then $y'_j(s) = y_j(s)$, $s \in [t_i - \delta^*, t_i + \delta^*]$. Otherwise $j_{i-1} = j_i = j$ are members of R such that either $t_{i-1} < t_i$ or $t_i < t_{i-1}$. We consider first the case $t_{i-1} < t_i$ and suppose that

$$y'_j(t_{i-1}-\delta^*) = y_j(t_{i-1}-\delta^*).$$

Let $m = j_{i-2}$; $m \neq j$ by (A) above. Then for almost every $s \in (t_{i-1}-\delta^*, t_{i-1}+\delta^*)$ either

$$x'_{jm}(s) = x_{jm}(s) - z^*,$$

or

$$x'_{mj}(s) = x_{mj}(s) + z^*,$$

and hence

$$\sum_{k=1}^{n} [x'_{kj}(s) - x'_{jk}(s)] = \sum_{k=1}^{n} [x_{kj}(s) - x_{jk}(s)] + z^*.$$

Thus

$$y'_j(s) = y_j(s) + \int_{t_{i-1}-\delta^*}^{s} z^* \, d\tau, \qquad s \in [t_{i-1}-\delta^*, t_{i-1}+\delta^*].$$

Since, for $s \in [t_{i-1}+\delta^*, t_i - \delta^*]$, both $x'_{jk}(s) = x_{jk}(s)$, and $x'_{kj}(s) = x_{kj}(s)$, we have

$$y'_j(s) - y_j(s) = 2z^*\delta^* = w^*, \qquad s \in [t_{i-1}+\delta^*, t_i - \delta^*].$$

Similarly, we can show

$$y'_j(s) = y_j(s) - \int_{t_i-\delta^*}^{s} z^* \, d\tau + w^*, \qquad s \in [t_i - \delta^*, t_i + \delta^*].$$

Thus

$$y'_j(s) - y_j(s) \leqslant w^*, \qquad s \in [t_{i-1}-\delta^*, t_i + \delta^*],$$

and $y'_j(t_i+\delta^*) = y_j(t_i+\delta^*)$.

Consider next the case $t_i < t_{i-1}$. Suppose that $y'_j(t_i - \delta^*) = y_j(t_i - \delta^*)$. Then, by an argument similar to that above, it can be shown that

$$y'_j(s) = y_j(s) - \int_{t_i-\delta^*}^{s} z^* d\tau, \qquad s \in [t_i - \delta^*, t_i + \delta^*],$$

$$y'_j(s) = y_j(s) - w^*, \qquad s \in [t_i + \delta^*, t_{i-1} - \delta^*],$$

$$y'_j(s) = y_j(s) + \int_{t_{i-1}-\delta^*}^{s} z^* \, d\tau - w^*, \qquad s \in [t_{i-1}-\delta^*, t_{i-1}+\delta^*].$$

By induction, it is clear from the argument above that y'_j differs from y_j only on intervals having endpoints $\min\{t_{i-1}, t_i\} - \delta^*$ and $\max\{t_{i-1}, t_i\} + \delta^*$, where i is such that $j_i = j_{i-1} = j$. Also, since $0 < w^* < w^{(i)}$ in both cases, y'_j is such that constraints (3) and (4) are satisfied. Thus $\{x'_{jk}\}$ is feasible with a positive improvement in flow of w^*.

Finally consider the case $t_1 = T$; here $y_{j_1}(T) > 0$ by rule (f). Since y_{j_1} is continuous, we can find positive $\varepsilon^{(1)}$and $w^{(1)}$ such that $y_{j_1}(s) \geqslant w^{(1)}$ for $s \in (T - 2\varepsilon^{(1)}, T)$. Using a similar argument to that above we can choose δ^* and z^* and construct an improved flow $\{x'_{jk}\}$ which is feasible but now has $y_{j_1}(T) = y_{j_1}(T) - w^*$. Hence the lemma is proved. □

Lemma 6.3 *Given a piecewise continuous feasible flow $\{x_{jk}(t)\}$, suppose that node n is not reachable for any $t \in [0, T]$. Let a generalized cut C be defined by*

$$C(t) = \{\, j : j \text{ is reachable at time } t\};$$

then the value of the flow is equal to the value of the cut.

Proof. Firstly, node 1 is reachable and node n is not, and it is easy to show that the set of times where each node is reachable is open in $[0, T]$. Thus C is well defined and for each node j, we can define the sets Γ_j, P_j, Q_j, and R_j as in Section 6.3. Let V be the value of the cut C. We begin by showing that V is finite.

If $V = \infty$, then for any $K > 0$ we can find m such that for some j

$$\sum_{i=1}^{m} a_j(\beta_{ij}) > K.$$

But from the definition of C, we have that

$$y_j(t) = a_j(t), \qquad t \in R_j,$$
$$y_j(t) = 0, \qquad t \in P_j,$$

otherwise node j would be reachable at these times, contrary to the definition of β_{ij} and α_{ij}. Thus

$$\begin{aligned}
\sum_{i=1}^{m} a_j(\beta_{ij}) &= \sum_{i=1}^{m} y_j(\beta_{ij}) \\
&= \sum_{i=1}^{m} [y_j(\beta_{ij}) - y_j(\alpha_{ij})] \\
&= \int_{\Gamma'_j} \sum_{k=1}^{n} [x_{kj}(\tau) - x_{jk}(\tau)]\, d\tau \\
&\leqslant T \sup_{0 \leqslant \tau \leqslant T} \left\{ \sum_{k=1}^{n} [b_{kj}(\tau) + b_{jk}(\tau)] \right\},
\end{aligned}$$

where

$$\Gamma'_j = \bigcup_{i=1}^{m} (\alpha_{ij}, \beta_{ij}).$$

This is a contradiction; thus V is finite and since $y_j(t)$ is non-negative for every j and t we can assume the existence of the infinite sums in the proof which follows.

Let v be the value of the flow $\{x_{jk}(t)\}$. Since $\Gamma_1 = (0, T)$ and $j \in \tilde{C}(T)$ implies

$y_j(T) = 0$, equation (9) of Section 6.3 becomes

$$v = \int_{\Gamma_1} \sum_{k=1}^{n} [x_{1k}(\tau) - x_{k1}(\tau)]\, d\tau - \sum_{j \in C(T)} y_j(T). \tag{10}$$

As $y_j(t) = 0$ for $t \in P_j$, we have from equations (7) and (10) and $\Gamma_n = \varnothing$,

$$v = \sum_{j=1}^{n} \int_{\Gamma_j} \sum_{k=1}^{n} [x_{jk}(\tau) - x_{kj}(\tau)]\, d\tau$$

$$+ \sum_{j=2}^{n-1} \sum_{i=1}^{\infty} y_j(\beta_{ij}) - \sum_{j \in C(T)} y_j(T).$$

It follows from the definition of R_j in Section 6.3 that

$$v = \sum_{j=1}^{n} \int_{\Gamma_j} \sum_{k=1}^{n} [x_{jk}(\tau) - x_{kj}(\tau)]\, d\tau + \sum_{j=2}^{n-1} \sum_{t \in R_j} a_j(t).$$

From the definition of Γ_j, it can be seen that

$$\sum_{j=1}^{n} \int_{\Gamma_j} \sum_{k=1}^{n} [x_{jk}(\tau) - x_{kj}(\tau)]\, d\tau = \sum_{j=1}^{n} \sum_{k=1}^{n} \int_{\Gamma_j} [x_{jk}(\tau) - x_{kj}(\tau)]\, d\tau$$

$$= \sum_{j=1}^{n} \sum_{k=1}^{n} \int_{\Gamma_j \cap \tilde{\Gamma}_k} [x_{jk}(\tau) - x_{kj}(\tau)]\, d\tau$$

$$= \sum_{j=1}^{n} \sum_{k=1}^{n} \int_{\Gamma_j \cap \tilde{\Gamma}_k} b_{jk}(\tau)\, d\tau.$$

This last step uses the piecewise continuity of x_{jk} and the fact that rules (b) and (c) have not made the nodes in $\tilde{\Gamma}_k$ reachable. Thus we have established that $v = V$.

□

Theorem 6.4 *If the maximum value v^* of the flow in the network Ω is achieved by a piecewise continuous flow, then it is equal to the minimum value V of all generalized cuts in Ω separating nodes* 1 *and* n.

Proof. Choose an optimal piecewise continuous feasible flow with value v^*; by Lemma 6.2, node n is not reachable at any time t. Thus we can construct a cut C defined by the reachable nodes, and for this cut, $v^* = V$ by Lemma 6.3. Since Lemma 6.1 ensures that every cut C' has value $V' \geqslant v^* = V$, the theorem is proved.

□

Ideally we should at this point like to establish that there will be a piecewise continuous optimal solution if the arc capacities are piecewise continuous and the storage capacities are sufficiently well behaved. However, no such result is available. We shall return to the question of piecewise continuity of optimal solutions when we consider the more general separated continuous linear program in Chapter 7. A result we give there (Theorem 7.6) shows that if the arc capacities are piecewise constant and the storage capacities are piecewise linear, then the optimal flows are piecewise constant. This is not very useful in this context, but seems to be the strongest result easily available.

In the next section we establish by a duality argument that, if we cannot find a cut with the same value as the optimal flow, we can at least find a sequence of cuts whose values approach that of the maximal flow.

In the above we have established continuous analogues of the results which form the theoretical basis of the classical labelling algorithm for the maximum flow problem. This analysis leads naturally to the specification of a continuous labelling algorithm which, given a trial flow, determines and labels those nodes that are reachable in specified subintervals of $(0, T)$. Such an algorithm has been implemented for networks having constant storage capacities, and arc capacities which are piecewise linear functions of time, by Philpott (1982). In this situation there are piecewise linear trial flows and quadratically varying storage.

A succession of passes is made over the network to label the reachable nodes according to the rules of Section 6.4. Failure to label the sink over any time interval then implies that an optimal solution has been reached. This can be checked by evaluating the cut defined by the reachable nodes. Otherwise, should node n become reachable during a labelling pass, we can increase the flow by Lemma 6.2. First, the flow augmenting path must be recovered. Following this, the maximum feasible increase in flow along this path is calculated.

6.5 DUALITY THEORY

The Ford–Fulkerson maximum-flow minimum-cut theorem has the appearance at least of a linear programming duality result. In this section we shall discuss the extent to which this is the case, first for the classical network-flow problem and then for the time-continuous version. This approach will enable us to establish a maximum-flow infimum-cut theorem.

We begin by giving the dual problem for the classical network-flow problem NP. The simplest way to write this down is as follows.

$$\begin{aligned}
\text{NP*:}\quad & \text{minimize} && \sum_{j=1}^{n}\sum_{k=1}^{n} b_{jk}\lambda_{jk} \\
& \text{subject to} && \pi_j - \pi_k + \lambda_{jk} - z_{jk} = 0, \qquad j, k = 1, 2, \ldots, n, \\
& && \pi_1 = 0, \qquad \pi_n = 1, \\
& && \lambda_{jk} \geqslant 0, \quad z_{jk} \geqslant 0, \qquad j, k = 1, 2, \ldots, n.
\end{aligned}$$

Here we have chosen to include slack variables z_{jk} to put the problem in equality-constrained form. At first sight this problem does not appear to be closely related to the minimum-cut problem, but we shall see below that the two problems are in fact equivalent.

As NP* is a finite linear program with a finite value, it achieves its minimum at an extreme point of the feasible set. For a solution (π, λ, z) to be an extreme point we must have

$$\lambda_{jk} z_{jk} = 0, \qquad j, k = 1, 2, \ldots, n, \tag{11}$$

$$\pi_j = 0 \text{ or } 1, \qquad j = 2, \ldots, n-1. \tag{12}$$

The first condition states that one or other of λ_{jk} and z_{jk} is zero. If they were both non-zero for some given j and k, then feasible solutions could be obtained with $\lambda'_{jk} = \lambda_{jk} \pm \varepsilon$, $z'_{jk} = z_{jk} \pm \varepsilon$, showing that the solution is not extreme.

The second condition follows from observing that if $\pi_j = p \in \{0,1\}$. for j in some set of indices J, and the other values of π lying outside $[p-\delta, p+\delta]$, then setting $\pi'_j = p \pm \delta, j \in J$ and making corresponding changes to either λ_{jk} or z_{jk} if just one of j and k is in J, again shows that the solution cannot be an extreme point.

As $\lambda_{jk} \geqslant 0$, equation (11) implies that

$$\lambda_{jk} = [\pi_k - \pi_j]_+.$$

Thus the problem NP* is equivalent to the minimum-cut problem:

$$\text{MCP: minimize} \quad \sum_{j=1}^{n} \sum_{k=1}^{n} b_{jk} [\pi_k - \pi_j]_+$$

$$\text{subject to} \quad \pi_j = 0 \text{ or } 1, \quad j = 2, \ldots, n-1,$$

$$\pi_1 = 0, \quad \pi_n = 1.$$

In attemping to establish the same kind of equivalence for the time-continuous network-flow problem, we need to begin by specifying the dual problem. To do this we shall choose different spaces in which to pose the problem than those assumed in Section 6.2 in order to show that the problem was solvable. The constraint map takes the primal variable space $L_\infty^{n^2}$ into $C^{n-2} \times C^{n-2} \times L_\infty^{n^2}$. In fact we shall take the primal constraint space Z to be $C^{n-2} \times C^{n-2} \times L_1^{n^2}$. Then a natural choice for the dual variable space W is $M_r^{n-2} \times M_r^{n-2} \times L_\infty^{n^2}$, where we write M_r for the space of regular signed Borel measures, and in all our notation we leave out mention of the interval $[0, T]$ on which the functions and measures are defined. Thus we obtain the following scheme

$$\text{TCNP:} \begin{bmatrix} L_\infty^{n^2} \\ L_1^{n^2} \end{bmatrix} \begin{matrix} \rightarrow \\ \leftarrow \end{matrix} \begin{bmatrix} C^{n-2} \times C^{n-2} \times L_1^{n^2} \\ M_r^{n-2} \times M_r^{n-2} \times L_\infty^{n^2} \end{bmatrix} \text{:TCNP*}.$$

With this choice of spaces the dual to the time-continuous network-flow problem takes the following form.

$$\text{TCNP*:} \quad \text{minimize} \quad \int_0^T \sum_{j=2}^{n-1} a_j(t)\, \mathrm{d}\lambda_j(t) + \int_0^T \sum_{j=1}^{n} \sum_{k=1}^{n} b_{jk}(t) \rho_{jk}(t)\, \mathrm{d}t$$

$$\text{subject to} \quad \int_t^T (\mathrm{d}\mu_j - \mathrm{d}\lambda_j - \mathrm{d}\mu_k + \mathrm{d}\lambda_k) + \rho_{jk}(t) \geqslant 0$$

$$\int_t^T (d\mu_j - d\lambda_j) + \rho_{j1}(t) \geqslant 0,$$

$$\int_t^T (d\mu_j - d\lambda_j) + \rho_{jn}(t) \geqslant 1,$$

$$\int_t^T (-d\mu_k + d\lambda_k) + \rho_{nk}(t) \geqslant -1,$$

$$\int_t^T (-d\mu_k + d\lambda_k) + \rho_{1k}(t) \geqslant 0,$$

$$j, k = 2, 3, \ldots, n-1,$$

$$\rho_{1n} \geqslant 1,$$

$$\mu_j \geqslant 0, \quad \lambda_j \geqslant 0, \quad \rho_{jk} \geqslant 0, \qquad j, k = 1, 2, \ldots, n.$$

This can be rewritten as

$$\text{minimize} \quad \int_0^T \sum_{j=2}^{n-1} a_j(t)\, d\lambda_j(t) + \int_0^T \sum_{j=1}^{n} \sum_{k=1}^{n} b_{jk}(t)\rho_{jk}(t)\, dt$$

$$\text{subject to} \quad \int_t^T (d\lambda_k - d\mu_k + d\mu_j - d\lambda_j) + \rho_{jk}(t) - \gamma_{jk}(t) = 0, \quad t \in [0, T] \quad (13)$$

$$\lambda_j \geqslant 0, \quad \mu_j \geqslant 0, \quad \rho_{jk} \geqslant 0, \quad \gamma_{jk} \geqslant 0, \qquad j, k = 1, 2, \ldots, n,$$

$$\lambda_1 = 0, \qquad \lambda_n = 0,$$

$$\mu_1 = 0, \qquad \mu_n \text{ assigns unit mass at } t = T.$$

We can apply Theorem 3.15 to show that there is no duality gap for TCNP and TCNP*, as the solution of TCNP is bounded by the b_{jk} constraints.

When the supports of μ_j and λ_j are finite for each j, we say that the solution (μ, λ, ρ) is *finite*.

Lemma 6.5 *If a finite feasible solution is basic, then the functions g_j, defined by*

$$g_j(t) = \int_t^T (d\lambda_j - d\mu_j), \qquad j = 1, 2, \ldots, n,$$

are 0 *or* -1 *almost everywhere, and* $\rho_{jk}(t) = \max(0, g_j(t) - g_k(t))$ *almost everywhere.*

Proof. The constraints ensure that

$$\rho_{jk}(t) \geqslant \max(0, g_j(t) - g_k(t)).$$

If there is strict inequality on a set I with positive measure, then there is a measurable set I' and $\varepsilon > 0$, with

$$\rho_{jk}(t) > \varepsilon + \max(0, g_j(t) - g_k(t))$$

on I'. Thus the solutions (μ, λ, ρ') with $\rho'_{jk}(t) = \rho_{jk}(t) \pm \varepsilon$, for t in I', and

$\rho'_{jk}(t) = \rho_{jk}(t)$ elsewhere, are feasible, and (μ, λ, ρ) is not basic. This establishes the second part of the lemma.

For the first part of the lemma, first note that $g_1(t) = 0$, $t \in [0, T]$, and $g_n(t) = -1$, $t \in [0, T)$. Now suppose that $g_j(t) = p$, for some t and some set of indices $j \in J$, with p neither 0 nor -1. For each $j \in J$ we may choose an interval (t_1, t_2) on which $g_j(t) = p$ and with one of μ_j and λ_j non-zero at t_1 and t_2, because the supports of μ_j and λ_j are finite. As t_1 and t_2 depend on j we write them as $t_1(j)$ and $t_2(j)$. Let

$$T_1 = \min\{t_1(j): j \in J), \qquad T_2 = \max\{t_2(j): j \in J\}.$$

Then we may define

$$\delta_1 = \min\{|g_k(t) - p|: g_k(t) \neq p, t \in (T_1, T_2)\},$$

and δ_2 to be the minimum of the masses which μ_j or λ_j places at $t_i(j)$, $i = 1, 2$, $j \in J$. Note that, since (μ, λ, ρ) is basic, in each case just one of these two masses is non-zero.

Now if δ is smaller than both δ_1 and δ_2, then we can construct new feasible solutions (μ', λ', ρ') with

$$g'_j(t) = g_j(t) \pm \delta, \qquad t \in (t_1(j), t_2(j)), \quad j \in J,$$
$$g'_j(t) = g_j(t), \qquad \text{otherwise.}$$

This is done by making the appropriate changes to the mass which either μ_j or λ_j has at $t_i(j)$, $i = 1, 2$, $j \in J$. As δ is less than δ_2 this can be done without making μ or λ become negative. Also ρ and γ are adjusted accordingly, and as δ is less than δ_1, both ρ and γ remain non-negative.

This contradicts the fact that (μ, λ, ρ) is basic and establishes the first part of the lemma. □

This result goes part of the way to establishing a correspondence between TCNP* and the minimum-cut problem. If the optimal solution to TCNP* is finite and basic, then it defines an optimal cut with

$$C(t) = \{i: g_i(s) = 0 \quad \text{for} \quad s \in (t - \delta, t + \delta) \text{ with some } \delta\}.$$

However, an optimal solution to TCNP* may not exist; moreover an optimal cut may have an infinite number of switches. Nevertheless, we can use this result to establish a maximum-flow infimum-cut theorem. As there is no duality gap for TCNP, this result can be proved by showing that there is a finite basic solution which gets as close as we like to the optimal value for the dual problem.

Theorem 6.6 *The value of the maximal flow for* TCNP *is equal to the infimum of the values of possible generalized cuts in* Ω.

Proof. Let v be the value of the maximal flow for TCNP. As there is no duality gap we may, for arbitrary $\varepsilon > 0$, choose a feasible solution (μ, λ, ρ) for TCNP* with an objective function value less than $v + \varepsilon$. The essential idea of the proof is to partition the interval $[0, T]$ using a set $S = \{0 = t_0, t_1, t_2, \ldots, t_N = T\}$, and then approximate the solution (μ, λ, ρ) with a finite basic solution (whose support is within S). We need a result from measure theory which states that for any function $f \in L_\infty [0, T]$ and any $\delta > 0$, there exists a step function ϕ, having a finite number of steps, and a set Z in $[0, T]$ with measure less than δ, such that $|f(x) - \phi(x)| < \delta$ for every $x \in [0, T]$ not in Z.

For a given δ we use this result to define step maps ρ'_{jk} within δ/n^2 of ρ_{jk} except on sets Z_{jk} of measure less than δ/n^2. The points at which any of ρ'_{jk}, $j, k = 1,2, \ldots, n$, have steps give a starting point for the set S. But we need to augment this set in two ways. First we must not include points in any Z_{jk}. Any such points, q_j, $j = 1,2, \ldots, L$, are replaced by additional points s_j and r_j, $j = 1,2, \ldots, L$, not in Z_{jk}, and chosen so that $q_j \in (s_j, r_j)$ and $\Sigma(r_j - s_j) < \delta$. Second we must include sufficient extra points, also not in any Z_{jk}, so that adjacent points of S are within δ of each other. Suppose that $S = \{t_i, i = 0, 1, \ldots, N\}$ has been defined in this way.

Now we define μ'_j and λ'_j in $M_r[0,1]$, $j = 1,2, \ldots, n$, with masses concentrated on points in S, as follows

$$\mu'_j(t_i) = \int_{t_i}^{t_{i+1}} \mathrm{d}\mu_j,$$

$$\lambda'_j(t_i) = \int_{t_i}^{t_{i+1}} \mathrm{d}\lambda_j, \qquad j = 1,2, \ldots, n, \quad i = 0, 1, \ldots, N-1.$$

The effect of this is to ensure that $(\mu'\ \lambda', \rho)$ satisfies the original constraints (13) for each $t \in S$. Moreover μ', λ' define step functions

$$\rho'_{jk}(t) = \max\left\{0, \int_t^T (\mathrm{d}\lambda'_k - \mathrm{d}\mu'_k + \mathrm{d}\mu'_j - \mathrm{d}\lambda'_j)\right\}, \qquad j, k = 1,2, \ldots, n.$$

Thus $\rho'' = \rho$ on S. So except on the sets Z_{jk} and the intervals (s_j, r_j), ρ''_{jk} are step functions which are within δ/n^2 of ρ'_{jk}, and hence within $2\delta/n^2$ of ρ_{jk}.

Now suppose that the modulus of $\int_0^T \mathrm{d}\lambda_j$, $\rho_j(t)$ and $b_{jk}(t)$, $j, k = 1,2, \ldots, n$ are all bounded by M, say. Write γ_δ for the maximum change in the modulus of any a_j between two points less than δ apart. Then the objective function values for (μ, λ, ρ) and (μ', λ', ρ'') differ by less than

$$n\gamma_\delta M + 2\delta MT + 4\delta n^2 M^2. \tag{14}$$

In this expression the first term comes from the term in the objective functional involving the a_j; the second term follows from observing that outside an exceptional set ρ and ρ'' are close to each other; while the third term results from the fact that this exceptional set is of measure at most 2δ and on it ρ and ρ'' differ by at most $2M$. It is clear from the continuity of the a_j that we may choose δ to make (14) as small as we like.

Of course (μ', λ', ρ'') may not be a basic solution. Consider solving TCNP* subject to the restriction that the supports of μ and λ must le within S. This is then a finite linear program and it has an optimal solution which is basic for TCNP* and which has an objective functional value no more than that for (μ', λ', ρ'').

Hence we can ensure that this solution corresponds to a cut which, by choosing δ and small enough, has a value within 2ε of v. □

6.6 NOTES

The original work on the network flow problem was done by Ford and Fulkerson (1962). Since then their maximum-flow minimum-cut theorem has appeared in innumerable textbooks and is one of the most famous theorems in elementary optimization theory.

One straightforward extension of the network-flow problem is to the case where a choice of flows has to be made at successive times, and the arc capacities vary with time. In the absence of storage within the network the solution is simple, in principle at least; one simply has to solve a static problem at each successive time. This problem has been discussed by Abel *et al.* (1979)

A non-trivial extension of the classical problem is the maximal dynamic flows problem formulated and solved by Ford and Fulkerson; here a choice of flows must be made at successive times, where each arc has a constant capacity and traversal time and unlimited holdovers (storage) are permitted at intermediate nodes. Halpern (1979) has considered the further extension to the case where arc capacities vary with time and storage may be prohibited at some times. When the time variable takes discrete values, both these problems can be solved by constructing equivalent large static problems. These constructions, which are described in Section 6.2, are still valid when upper bounds are imposed on the storage at each node.

The material on the time-continuous network-flow problem in this chapter is taken for the most part from Anderson, Nash and Philpott (1982) and Philpott (1982). The example given at the end of Section 6.3 is due to Ogier (1985), who thus shows that one of the results given in Anderson *et al.* (1982) is incorrect. Ogier (1985) has also proved a version of Theorem 6.6 using an entirely different approach.

Problems

6.1 Consider the maximal dynamic-flow problem when there is a transmission delay, d_{jk}, associated with each arc (j, k). Thus the constraints (3) and (4) of TCNP become

$$0 \leqslant \int_0^t \sum_{k=1}^{n} (x_{kj}(\tau - d_{kj}) - x_{jk}(\tau))\,\mathrm{d}\tau \leqslant a_j(t)$$

for $j = 2, 3, \ldots, n$ $t \in [0, T]$. Show how the discrete-time version of this problem can be formulated as a maximum-flow problem in a time-augmented network. Hence define the value of a cut in this situation and prove a version of Theorem 6.1.

6.2 Construct an example of a problem in the form TCNP with continuous arc capacities, but with a minimal cut having an infinite number of open intervals for some Γ_j.

6.3 Prove the statement made in the proof of Lemma 6.3 that, for any node, the set of times for which it is reachable is an open set.

6.4 Derive the forms of the dual problems NP* and TCNP* given in Section 6.5.

6.5 Find a sequence of cuts whose value approaches $m(R)$ for the example given at the end of Section 6.3.

6.6 Consider the network defined on the nodes $\{1,2,3,4\}$ over the time interval $[0,1]$, with arc capacities

$$b_{12}(t) = 1, \quad b_{13}(t) = 4, \qquad t \in [0, 1/4],$$
$$b_{23}(t) = b_{32}(t) = b_{24}(t) = b_{34}(t) = t, \qquad t \in [0,1],$$

and all other arc capacities zero; and with

$$a_2(t) = t/2, \quad a_3(t) = 1, \qquad t \in [0, 1].$$

Find a maximal flow and a minimal cut.

7

Continuous Linear Programs

7.1 INTRODUCTION

In this chapter we turn our attention to possibly the most important application of infinite-dimensional linear programming: the linear optimal control problem with linear state constraints. Such problems are often called continuous linear programs, and occur in a variety of contexts. In practice they are usually solved by discretizing over time to give a large finite linear program. However, the possibility of obtaining an optimal solution more directly, by linear programming methods, has acted as a spur to research over many years. Unfortunately, relatively little progress has been made in this direction.

We shall start by surveying the duality theory of continuous linear programs and then look in detail at a more restricted class of problems, called separated continuous linear programs.

7.2 THE CONTINUOUS LINEAR PROGRAM AND ITS DUAL

The bottleneck problem was introduced in Section 1.2.1. Such problems are called *continuous linear programs*; they have the following form:

$$\text{CLP: maximize} \quad \int_0^T c(t)^{\mathrm{T}} x(t)\,\mathrm{d}t$$

$$\text{subject to} \quad Bx(t) - \int_0^t Kx(s)\,\mathrm{d}s \leqslant b(t),$$

$$x(t) \geqslant 0, \qquad t \in [0, T].$$

Here $x(t)$, $c(t) \in R^n$, $b(t) \in R^m$ and B and K are $m \times n$ matrices.

We have chosen to pose this problem as a maximization, rather than as a minimization, because that is standard in the literature. On the other hand we have taken B and K to be constant, where the most common formulation has constraints

$$B(t)x(t) - \int_0^t K(s, t)x(s)\,\mathrm{d}s \leqslant b(t).$$

In fact the development that we give can be carried through largely unchanged for this form of problem.

In order to put the program into a topological framework, we take the primal and dual variable spaces to be $L^n_\infty[0,T]$ and the primal and dual constraint spaces to be $L^m_1[0,T]$:

$$\text{CLP:}\begin{bmatrix} L^n_\infty[0,T] \\ L^n_1[0,T] \end{bmatrix} \begin{matrix} \rightarrow \\ \leftarrow \end{matrix} \begin{bmatrix} L^m_1[0,T] \\ L^m_\infty[0,T] \end{bmatrix}\text{:CLP*}$$

There are two reasons for choosing this particular topological setting for the continuous linear program. First, the main interest in the duality theory of CLP has been with the case when both it and its dual are posed over L_∞ spaces. Secondly, by making this choice of spaces we can satisfy the conditions of Theorem 3.15, which we shall use later to establish a duality theorem.

We need to check that the constraint map is continuous in the appropriate weak topologies. This can be done by showing that the adjoint of the constraint map takes elements of L_∞ to L_1. We write A for the constraint map, so that

$$(Ax)(t) = Bx(t) - \int_0^t Kx(s)\,\mathrm{d}s.$$

We wish to find A^* with $\langle Ax, w\rangle = \langle x, A^*w\rangle$. Now

$$\begin{aligned}\langle Ax, w\rangle &= \int_0^T (Bx(t))^{\mathrm{T}}w(t)\,\mathrm{d}t - \int_0^T w(t)^{\mathrm{T}}\left(\int_0^t Kx(s)\,\mathrm{d}s\right)\mathrm{d}t \\ &= \int_0^T x(t)^T(B^{\mathrm{T}}w(t))\,\mathrm{d}t - \int_0^T x(t)^{\mathrm{T}}\left(\int_t^T K^{\mathrm{T}}w(s)\,\mathrm{d}s\right)\mathrm{d}t\end{aligned}$$

using Fubini's theorem. So

$$(A^*w)(t) = B^{\mathrm{T}}w(t) - \int_t^T K^{\mathrm{T}}w(s)\,\mathrm{d}s.$$

Thus the adjoint map takes L_∞ into L_1 (actually into L_∞) and A is continuous in the appropriate weak topologies. Having found the adjoint of A, we can write down the dual problem

$$\begin{aligned}\text{CLP*:}\quad &\text{minimize} && \int_0^T b(t)^{\mathrm{T}}w(t)\,\mathrm{d}t \\ &\text{subject to} && B^{\mathrm{T}}w(t) + \int_t^T K^{\mathrm{T}}w(s)\,\mathrm{d}s \geqslant c(t), \qquad t\in[0,T], \\ & && w(t) \geqslant 0.\end{aligned}$$

In the remainder of this section we shall give a set of conditions which ensure that there is no duality gap for CLP and its dual. Such conditions are quite difficult to find, as the reader will appreciate from our discussion of the example P0 in Chapter 3. Before starting our discussion of duality theory, we shall prove two preliminary lemmas. The first of these is concerned with convex sets in finite-dimensional spaces and is a well-known result (see for example Holmes (1975)).

For a convex set B in R^n we define the *recession cone* C_B as follows:

$$C_B = \{x \in R^n: x + B \subset B\}.$$

It is not difficult to see that this is a convex cone, for if x and y are in C_B and $\lambda \geq 0$, then for any $z \in B$,

$$\lambda(x+y)+z = \lambda(x+(y+z))+(1-\lambda)z,$$

which is a convex combination of two points in B and hence is in B, so $\lambda(x+y)$ is in C_B.

If B is closed the recession cone consists of those directions in which B is unbounded, these being the same for each point in B. For suppose that $x \in B$ with $x+\lambda y \in B$ for all $\lambda \geq 0$, then for any z in B, define

$$w_n = (n-1)z/n + (x+ny)/n.$$

Then $w_n \in B$ and w_n approaches $z+y$ as n tends to infinity. So $z+y \in B$ and hence $y \in C_B$.

Lemma 7.1 *If B is a closed convex set in R^n which does not contain any line, then for every x in B we can write $x = x_1 + x_2$ with x_1 in the convex hull of the extreme points of B and x_2 in C_B.*

Proof. The proof is by induction on the dimension of the space. Certainly the theorem holds when $n = 1$. We suppose that it holds for $n = m$ and take $n = m+1$. For x in B choose an arbitrary line through x. There are two possibilities:

(a) The line intersects B in a closed half line with endpoint z. Since z is on the boundary of B, there is a supporting hyperplane to B at z, say H. In $H \cap B$ we can apply the induction hypothesis and $z = z_1 + z_2$ with z_1 in the convex hull of the extreme points of $H \cap B$ and z_2 in $C_{H \cap B}$. Now the extreme points of $H \cap B$ are also extreme points of B. Moreover as B is closed the recession cone $C_{H \cap B}$ is contained in C_B. Thus $x = z_1 + z_2 + w$, say, where $x + \lambda w \in B$ for any $\lambda \geq 0$. This implies that w is in C_B and so $z_2 + w$ is also in C_B. Hence the inductive hypothesis is established.

(b) The line intersects B in a closed line segment with end points z_1 and z_2. Using the same arguments as in part (a), we can express both z_1 and z_2 as the sum of a point in the convex hull of the extreme points of B and a point in the recession cone C_B. As both these sets are convex and x is a convex combination of z_1 and z_2, we again establish the inductive hypothesis. □

The second lemma that we need gives a bound on the size of any function satisfying a certain integral inequality constraint.

Lemma 7.2 (Gronwall) *If $g(t)$ is an integrable function on $[0, T]$ and there exist positive constants α and β with*

$$g(t) \leqslant \alpha + \beta \int_0^t g(s)\,ds \qquad \textit{for almost all } t \in [0, T], \tag{1}$$

then

$$g(t) \leqslant \alpha e^{\beta t} \textit{ for almost all } t \in [0, T].$$

Proof. Let

$$G(t) = \int_0^t g(s)\,ds.$$

Then using the inequality (1)

$$\frac{d}{dt}(e^{-\beta t}G(t)) = e^{-\beta t}g(t) - \beta e^{-\beta t}G(t)$$

$$\leqslant \alpha e^{-\beta t} + \beta e^{-\beta t}\int_0^t g(s)\,ds - \beta e^{-\beta t}G(t),$$

$$\text{for almost all } t \in [0, T] = \alpha e^{-\beta t}.$$

Thus

$$e^{-\beta t}G(t) \leqslant \int_0^t \alpha e^{-\beta s} ds$$

$$= \alpha(1 - e^{-\beta t})/\beta,$$

so

$$G(t) \leqslant \alpha(e^{\beta t} - 1)/\beta.$$

Using the inequality (1) again we obtain the result. □

Now we can proceed to establish conditions under which strong duality holds for CLP, using Theorem 3.15. We need to introduce some notation. If $y \in R^m$ and A is an $n \times m$ matrix, we let

$$L(A, y) = \{x \in R^n\colon Ax \leqslant y,\ x \geqslant 0\}.$$

The recession cone of $L(A,y)$ consists of points $x \in R^n$ with $x \geqslant 0$ and $Ax \leqslant 0$. Write $H(A, y)$ for the convex hull of the extreme points of $L(A, y)$. Thus Lemma 7.1 implies that for each x in $L(A, y)$, $x = x_1 + x_2$ with $x_1 \in H(A, y)$, $x_2 \geqslant 0$ and $Ax_2 \leqslant 0$. Grinold(1968, Appendix D) has shown that if $x(t)$ and $y(t)$ are measurable and satisfy $x(t) \in L(A, y(t))$, $t \in [0, T]$, then $x(t) = x_1(t) + x_2(t)$ with $x_1(t) \in H(A, y(t))$, $x_2(t) \geqslant 0$, $Ax_2(t) \leqslant 0$ and both x_1 and x_2 measurable. This apparently simple remark is quite difficult to establish, and we shall not include a proof of this result here.

We introduce the following two conditions on CLP.

(A) There is some $\rho \geqslant 0$ such that for every d for which $L(B, d)$ is non-empty, $\|H(B, d)\| \leqslant \rho \|d\|$,

(B) If $z \geqslant 0$ and $Bz \leqslant 0$, then both $c(t)^{\mathrm{T}}z \leqslant 0$, $t \in [0, T]$ and $Kz \leqslant 0$.

Theorem 7.3 *If* CLP *has a finite value and conditions* (*A*) *and* (*B*) *hold, then the values of* CLP *and* CLP* *are the same and* CLP *is solvable.*

Proof. We shall use Theorem 3.15 to prove this result. To do that, we must find a subset of the feasible region, depending on the function b, which is bounded by a multiple of $\|b\|$ and which contains an optimal solution.

Suppose that $x(t)$, $t \in [0, T]$ is feasible for CLP(b). Let

$$d(t) = b(t) + \int_0^t Kx(s)\,\mathrm{d}s.$$

Thus $x(t) \in L(B, d(t))$. Then from the remarks above we can find measurable functions x_1 and z_1, such that $x(t) = x_1(t) + z_1(t)$ and

$$x_1(t) \in H(B, d(t)), \quad z_1(t) \geqslant 0, \quad Bz_1(t) \leqslant 0.$$

From condition (B) we have $Kz_1(t) \leqslant 0$. Thus if we define $d_1(t)$ by

$$d_1(t) = b(t) + \int_0^t Kx_1(s)\,\mathrm{d}s,$$

we have $d(t) \leqslant d_1(t)$, and so $x_1(t) \in L(B, d_1(t))$.

We can repeat this process, obtaining a sequence of measurable functions x_n, d_n with

$$x_{n-1}(t) = x_n(t) + z_n(t),$$

$$x_n(t) \in H(B, d_{n-1}(t)), \qquad z_n \geqslant 0, \quad Kz_n(t) \geqslant 0,$$

$$d_n(t) = b(t) + \int_0^t Kx_n(s)\,\mathrm{d}s,$$

$$x_n(t) \in L(B, d_n(t)).$$

As the sequence $\{x_n(t)\}$ decreases monotonically, it has a limit, and we can define the measurable function x' by

$$x'(t) = \lim_{n \to \infty} x_n(t).$$

Moreover, $0 \leqslant x_n(t) \leqslant x(t)$ and the integral

$$\int_0^t Kx(s)\,\mathrm{d}s$$

is finite. So, by Lebesgue's dominated convergence theorem, Kx' is integrable and

$$\lim_{n \to \infty} d_n(t) = b(t) + \int_0^t Kx'(s)\,\mathrm{d}s.$$

We write d' for this function. The extreme points of $L(B, d_n(t))$ approach those of $L(B, d'(t))$ and this is sufficient to show that $x'(t) \in H(B, d'(t))$.

From condition (A),

$$\|x'(t)\| \leqslant \rho \|d'(t)\|$$

$$\leqslant \rho\|b(t)\| + \rho\left\|\int_0^t Kx'(s)\,\mathrm{d}s\right\|$$

$$\leqslant \rho\|b\| + \rho\alpha\int_0^t \|x'(s)\|\,\mathrm{d}s, \qquad t\in[0,T],$$

where $\|b\|$ is the L_∞ norm of b and α is the norm of K. Hence, using Gronwall's lemma,

$$\|x'(t)\| \leqslant \rho\|b\|\mathrm{e}^{\rho\alpha t}, \qquad \text{for almost all } t\in[0,T],$$

and in the L_∞ norm

$$\|x'\| \leqslant \rho\mathrm{e}^{\rho\alpha T}\|b\|.$$

Now

$$x(t) - x'(t) = \lim_{n\to\infty}\sum_{i=1}^{n} z_i(t),$$

so, using condition (**B**),

$$c(t)^{\mathrm{T}}x(t) - c(t)^{\mathrm{T}}x'(t) = \lim_{n\to\infty}\sum_{i=1}^{n} c(t)^{\mathrm{T}}z_i(t) \leqslant 0.$$

Thus we have all the conditions necessary to apply Theorem 3.15 to show that there is no duality gap. The solvability of CLP follows immediately from Theorem 3.22. □

The conditions of this theorem are very restrictive, but there do not appear to be any stronger results available.

7.3 SEPARATED CONTINUOUS LINEAR PROGRAMS (SCLP)

In the remainder of this chapter, we consider a subclass of continuous linear programming problems, a member of which is called a *separated continuous linear program* (SCLP). The form of SCLP is as follows:

$$\text{SCLP: maximize} \quad \int_0^T c(t)^{\mathrm{T}}x(t)\,\mathrm{d}t$$

$$\text{subject to} \quad \int_0^t Gx(s)\,\mathrm{d}s + y(t) = a(t), \tag{2}$$

$$Hx(t) + z(t) = b(t), \tag{3}$$

$$x(t),\quad y(t),\quad z(t) \geqslant 0, \qquad t\in[0,T]. \tag{4}$$

Here x, z, b and c are (vector-valued) functions in $L_\infty[0,T]$; y and a are absolutely continuous functions. The dimensions of x, y and z are n_1, n_2 and n_3 respectively, the dimensions of a, b and c are n_2, n_3 and n_1 respectively. We set $n = n_1 + n_2 + n_3$ and $m = n_2 + n_3$. The functions y and z are just slacks; it is convenient to pose the problem in equality-constrained form from the outset. The description 'separated' refers to the fact that the constraints are in two sets, the integral constraints (2) and the instantaneous constraints (3) and (4).

The separated continuous linear program has an alternative formulation as a linear optimal control problem, with state positivity constraints:

$$\begin{aligned}
&\text{maximize} && \int_0^T \{\alpha(t)^{\mathrm{T}} y(t) + \beta(t)^{\mathrm{T}} u(t)\}\,\mathrm{d}t \\
&\text{subject to} && \dot{y}(t) = h(t) - Gu(t), \\
& && y(0) = a(0), \\
& && u(t) \in U(t) = \{u : u \geqslant 0,\ Hu \leqslant b(t)\}, \\
& && y(t) \geqslant 0, \quad t \in [0, T].
\end{aligned}$$

The form of SCLP can be recovered from this by formally integrating the dynamics and removing the state dependence from the objective by integrating by parts.

Problems of this type arise in considering continuous-time, multi-commodity network-flow problems of various kinds, including production and inventory problems, deterministic reservoir control problems, and transportation and storage problems. Indeed we have already met a separated continuous linear program in the time-continuous network-flow problem of Chapter 6. The integral constraints (2) represent constraints on storage at nodes of the network, for example stockpiles or reservoirs, while the instantaneous constraints (3) typically represent constraints on processing or transportation capacity in the arcs. In the final chapter we shall look in more detail at infinite-dimensional problems specifically related to minimizing the cost of flow in a network.

Taking L_∞ for the dual variable and constraint spaces, the dual of SCLP is

$$\begin{aligned}
&\text{SCLP*:}\quad \text{minimize} && \int_0^T (a(t)^{\mathrm{T}} u(t) + b(t)^{\mathrm{T}} v(t))\,\mathrm{d}t \\
&\text{subject to} && \int_t^T G^{\mathrm{T}} u(s)\,\mathrm{d}s + H^{\mathrm{T}} v(t) - w(t) = c(t), \\
& && u(t) \geqslant 0, \quad v(t) \geqslant 0, \quad w(t) \geqslant 0, \qquad t \in [0, T].
\end{aligned}$$

In the next section we shall discuss in detail a characterization of the basic solutions for the separated continuous linear program. Before doing so we point out that the separated continuous linear program is solvable if the feasible region is bounded in the L_∞ norm. This follows immediately from Theorem 3.20.

7.4 THE BASIC SOLUTIONS FOR SCLP

For any x in $L_\infty^n[0, T]$ define the *support* of x, denoted by S_x, to be a set-valued function on $[0, T]$ such that $S_x(t)$ is the set of indices of non-zero components of

$x(t)$; that is

$$S_x(t) = \{k: |x_k(t)| > 0\}, \quad t\in[0, T].$$

Strictly, of course, this defines an equivalence class of set-valued functions differing from each other on sets of measure zero. This has no essential bearing on what follows, and we will ignore it for the sake of brevity in the exposition.

Define

$$\hat{x}(t) = \begin{bmatrix} x(t) \\ y(t) \\ z(t) \end{bmatrix}.$$

Note that if $\hat{x}$ is feasible (i.e. if its components x, y and z satisfy (2), (3) and (4)), then it is determined from x alone. Define the $n \times m$ matrix K by

$$K = \begin{bmatrix} G & I & 0 \\ H & 0 & I \end{bmatrix}.$$

Note that rank $(K) = m$. Equation (2) can be differentiated, at least almost everywhere, to give

$$Gx(t) + \dot{y}(t) = \dot{a}(t). \tag{5}$$

Let

$$\bar{x}(t) = \begin{bmatrix} x(t) \\ \dot{y}(t) \\ z(t) \end{bmatrix}.$$

Again, if $\hat{x}$ is feasible, then $\bar{x}$ is determined from x alone. Clearly $\bar{x}$ will also determine $\hat{x}$. Equations (2) and (3) are equivalent to

$$K\bar{x}(t) = \begin{bmatrix} \dot{a}(t) \\ b(t) \end{bmatrix}, \tag{6}$$

$$y(0) = a(0).$$

Throughout what follows we shall assume that $a(t)$ is differentiable everywhere and $\dot{a}(t)$ is bounded. We can now give a characterization of the basic solutions for SCLP.

Theorem 7.4 *Suppose that $\hat{x}$ is feasible for* SCLP. *Then $\hat{x}$ is basic if and only if the columns of K indexed by $S_{\hat{x}}(t)$ are linearly independent for almost all t in* $[0, T]$.

Proof. Suppose first that $\hat{x}$ is an extreme point of the feasible set, but the columns of K indexed by $S_{\hat{x}}(t)$ are not linearly independent for almost all t in $[0, T]$. Let P be a set of non-zero measure on which the columns of K indexed by $S_{\hat{x}}(t)$ are linearly dependent. Choose some subset of P, say P', also of non-zero measure, on which $S_{\hat{x}}(t)$ is constant, say equal to S. If S does not contain the index of any y variable, we can use a standard argument from finite linear programming at each $t \in P'$ to produce a function which can be added and subtracted from $\hat{x}$ at each point in P' while maintaining feasibility, which is a contradiction. The difficulty is that when S does contain the index of some y variables, we have to show that this can be done without any of those variables becoming negative. This is done as follows.

By definition, the y variables indexed by S are strictly positive at each t in P'. Since the y variables are continuous functions, we can choose an $\varepsilon(t) > 0$ and an open interval I_t for each t in P', such that

$$\hat{x}_i(t') > \varepsilon(t) \text{ for all } i \in S,\ n_1 < i \leqslant n_1 + n_2, \text{ for all } t' \in I_t.$$

Let

$$J = \bigcup_{t \in P'} I_t.$$

Then $P' \subset J$, and so for at least one of the pairs $\{(I_t, \varepsilon(t))\}$, say (I, ε), $P'' = P' \cap I$ has a measure greater than zero, and

$$\hat{x}_i(t') > \varepsilon \text{ for all } i \in S,\ n_1 < i \leqslant n_1 + n_2, \text{ for all } t' \in I.$$

From the definition of P there is some $q \in R^n$, $q \neq 0$ with $Kq = 0$ and $q_i = 0$ for all $i \notin S$. If $\bar{x}'$ is defined from $\bar{x}$ by

$$\bar{x}'(t) = \bar{x}(t) + h(t)q, \quad t \in [0, T], \tag{7}$$

where $h(t)$ is any measurable scalar function, then $\bar{x}'$ satisfies (6). Hence $\hat{x}'$ (defined in the obvious way from $\bar{x}'$) is feasible if it is non-negative. We construct a function h which is zero outside P'' and such that $\hat{x}'$ is non-negative.

First define functions f_1 and f_2 by

$$f_1(t) = \min_{k \in I_1} \{\bar{x}_k(t)/q_k\}, \qquad I_1 \neq \varnothing,$$

$$f_2(t) = \min_{k \in I_2} \{-\bar{x}_k(t)/q_k\}, \qquad I_2 \neq \varnothing,$$

for each $t \in P''$, where

$$I_1 = \{k : q_k > 0, \quad k \leqslant n_1 \text{ or } k > n_1 + n_2\},$$

$$I_2 = \{k : q_k < 0, \quad k \leqslant n_1 \text{ or } k > n_1 + n_2\}.$$

If either I_1 or I_2 is empty, then the corresponding f_i is set to 1. Then f_1 and f_2 are bounded, measurable functions, non-zero on at least a subset of P'' of non-zero measure; f_1 (f_2) gives the instantaneous largest multiple of q which can be subtracted from (added to) $\bar{x}$ without making any component of x or z negative (except perhaps on a set of measure zero). Note that I_1 and I_2 cannot both be empty. Set $f(t) = \min(f_1(t), f_2(t))$, $t \in P''$. We can now choose disjoint sets P_1 and P_2 in P'', each with non-zero measure, and such that

$$\int_{P_1} f(t)\,\mathrm{d}t = \int_{P_2} f(t)\,\mathrm{d}t,$$

$$\left|q_i \int_{P_j} f(t)\,\mathrm{d}t\right| < \varepsilon, \qquad n_1 < i \leqslant n_1 + n_2, \quad j = 1,2.$$

Define $\bar{x}'$ from (7) by setting

$$\begin{aligned} h(t) &= f(t), && t \in P_1, \\ &= -f(t), && t \in P_2, \\ &= 0, && \text{otherwise,} \end{aligned}$$

and define $\hat{x}'$ from $\bar{x}'$ in the usual way. From the definition of f, x' and z' are non-negative. Also $\bar{x}$ is changed on a set which is small enough for y' to remain positive in I, and outside this interval y' is the same as y. Thus $\hat{x}'$ is non-negative and hence feasible. Furthermore, if we define $\hat{x}''$ in the same way, but setting

$$\begin{aligned} h(t) &= -f(t), && t \in P_1, \\ &= f(t), && t \in P_2, \\ &= 0, && \text{otherwise,} \end{aligned}$$

then this is also feasible. Moreover $\hat{x} = (\hat{x}' + \hat{x}'')/2$, which is a contradiction, as $\hat{x}$ is an extreme point of the set of feasible solutions by assumption.

Now suppose that the columns of K indexed by $S_{\hat{x}}(t)$ are linearly independent for almost all t in $[0, T]$ for some feasible $\hat{x}$ which is not an extreme point. Then there is some $\hat{x}'$ and $\hat{x}''$ in F with $\hat{x}$ an interior point of the line segment joining them. Define

$$g(t) = \bar{x}'(t) - \bar{x}''(t), \quad t \in [0, T].$$

Then

$$Kg(t) = K\bar{x}'(t) - K\bar{x}''(t) = 0, \quad t \in [0, T].$$

But

$$S_{\hat{x}}(t) = S_{\hat{x}'}(t) \cup S_{\hat{x}''}(t),$$

so $g_i(t)$ is zero if $i \notin S_{\hat{x}}(t)$, and $g(t)$ is non-zero on a set of measure greater than zero. This contradicts the assumption that the columns of K indexed by $S_{\hat{x}}(t)$ are linearly independent for almost all t in $[0, T]$. □

7.5 A CLASS OF SCLPs WITH PIECEWISE LINEAR SOLUTIONS

A common feature of SCLPs which arises in practice is that they have an optimal solution with the property that $S_{\hat{x}}$ is piecewise constant and has only a finite number of changes. An SCLP which has an optimal solution of this type is called *regional*. The behaviour of an optimal solution depends on the functions a, b and c, and it appears difficult to find general conditions on these functions under which an SCLP is regional. A rather restricted class of SCLPs which can be shown to be regional is discussed below. Moreover, for this class of problems $\hat{x}$ is linear on intervals where $S_{\hat{x}}$ is constant, giving optimal solutions which are piecewise linear. We assume throughout this section that a basic optimal solution exists.

Theorem 7.5 *If b is constant, and*

$$a(t) = a_1 + a_2 t, \qquad t \in [0, T],$$
$$c(t) = c_1 + c_2 t, \qquad t \in [0, T],$$

where a_1, a_2, c_1, c_2 are constant vectors, then SCLP has an optimal solution $\hat{x}$ with

$$\hat{c}_2^{\mathrm{T}} \hat{x}(t_1) \leqslant \hat{c}_2^{\mathrm{T}} \hat{x}(t_2), \tag{8}$$

for almost all $t_1 < t_2$, $t_1, t_2 \in [0, T]$, where

$$\hat{c}_2 = \begin{bmatrix} c_2 \\ 0 \\ 0 \end{bmatrix}.$$

Proof. Suppose that $\hat{x}$ is an optimal basic solution. Define $q(t)$ by

$$q(t) = \int_0^t \hat{c}_2^{\mathrm{T}} \hat{x}(s)\,\mathrm{d}s.$$

The proof proceeds by showing that q is a convex function. Suppose otherwise; then as q is continuous, there is some interval (t_1, t_2) such that

$$q(t) > r(t), \qquad t \in (t_1, t_2),$$

where

$$r(t) = q(t_1) + \frac{(t - t_1)}{(t_2 - t_1)}(q(t_2) - q(t_1)), \qquad t \in (t_1, t_2).$$

Now define x' by

$$x'(t) = (t_2 - t_1)^{-1} \int_{t_1}^{t_2} x(s)\,\mathrm{d}s, \qquad t \in (t_1, t_2),$$
$$x'(t) = x(t), \qquad \text{otherwise.}$$

Define $\hat{x}'$ from x' in the usual way. Then $\hat{x}'$ is feasible. Comparing objective functional values for $\hat{x}$ and $\hat{x}'$ gives

$$\begin{aligned}
&\int_0^T \hat{c}(s)^{\mathrm{T}} \hat{x}'(s)\,\mathrm{d}s - \int_0^T \hat{c}(s)^{\mathrm{T}} \hat{x}(s)\,\mathrm{d}s \\
&= \int_{t_1}^{t_2} s c_2^{\mathrm{T}} (x'(s) - x(s))\mathrm{d}s \\
&= \int_{t_1}^{t_2} s \frac{\mathrm{d}}{\mathrm{d}s}(r(s) - q(s))\mathrm{d}s \\
&= \int_{t_1}^{t_2} (q(s) - r(s))\,\mathrm{d}s - t_2(r(t_2) - r(t_1) - q(t_2) + q(t_1)) \\
&> 0,
\end{aligned}$$

since $r(t_1) = q(t_1)$, $r(t_2) = q(t_2)$ and $r(s) < q(s)$ on (t_1, t_2). This contradicts the assumed optimality of x. Hence $q(t)$ is convex and so its derivative is monotonic increasing (see, for example, Rockafellar (1970)). But its derivative is equal to $\hat{c}_2^{\mathrm{T}} \hat{x}(t)$ almost everywhere, which establishes the theorem. □

Theorem 7.6 *Under the conditions of Theorem 7.5, there is an optimal solution for which $\hat{x}$ is a piecewise linear function.*

Proof. The choice of support for a basic solution uniquely defines $\bar{x}(t)$ from (6), and hence, because of the form of $\hat{c}_2$, it determines $\hat{c}_2^{\mathrm{T}} \hat{x}(t)$. Suppose that the number of possible choices of $S_{\hat{x}}(t)$ is L (which is finite). Moreover, given a particular choice of $S_{\hat{x}}$, (6) determines the same value of $\bar{x}(t)$ for all t. Denote the possible values of $\bar{x}$ by $\bar{x}^{(i)}$, $i = 1, 2, \ldots, L$.

Suppose first that

$$\hat{c}_2^{\mathrm{T}} \hat{x}^{(i)} \neq \hat{c}_2^{\mathrm{T}} \hat{x}^{(j)}, \qquad i \neq j.$$

Without loss of generality, we may take

$$\hat{c}_2^{\mathrm{T}} \hat{x}^{(i)} < \hat{c}_2^{\mathrm{T}} \hat{x}^{(j)}, \qquad i < j, \tag{9}$$

where $\hat{x}^{(i)}$ is defined from $\bar{x}^{(i)}$, $i = 1, 2, \ldots, L$. Then the result follows immediately from Theorem 7.5.

Now consider the case when (9) does not hold. Choose a sequence $\{c_2^{(n)}\}$ approaching c_2, and such that (9) does hold if c_2 is replaced by any $c_2^{(n)}$ in the definition of $\hat{c}_2$. Let $\hat{x}(t)^{(n)}$ be an optimal solution to the problem with c_2 replaced by $c_2^{(n)}$. From the above, $\hat{x}(t)^{(n)}$ can be chosen to satisfy the conditions of the theorem. Thus $\hat{x}(t)^{(n)}$ can be described by a vector $p^{(n)}$ in R^L, with

$$\bar{x}(t)^{(n)} = \bar{x}^{(i)}, \qquad \text{for almost all } t \in \left(\sum_{j=0}^{i-1} p_j^{(n)}, \sum_{j=0}^{i} p_j^{(n)} \right],$$

and $\hat{x}^{(n)}$ defined from $\bar{x}^{(n)}$ in the usual way. As each component of $p^{(n)}$ is bounded by T, we may choose a subsequence from $p^{(n)}$ converging to p, say. Define $\bar{x}(t)$ by

$$\bar{x}(t) = \bar{x}^{(i)}, \qquad t \in \left(\sum_{j=0}^{i-1} p_j, \sum_{j=0}^{i} p_j \right].$$

Then it is clear that $\hat{x}$ defined from $\bar{x}$ in the usual way is optimal for the original problem, and the theorem is proved. □

Theorems 7.5 and 7.6 can be extended to problems in which a and c are piecewise linear, and b is piecewise constant, with a finite number of breakpoints. Suppose that $\{t_1, t_2, \ldots, t_{r-1}\}$ is the set of times at which a, b or c has a break point, with $t_0 = 0$, $t_r = T$. Then the proofs of Theorems 7.5 and 7.6 hold when applied between t_{i-1} and t_i, $i = 1, 2, \ldots, r$.

Define a *region* (t_1, t_2) to be an interval over which the support of $\hat{x}$ is constant. It is worth pointing out that Theorem 7.6 allows a separated continuous linear program to be solved as a quadratic program. This can be done by choosing as variables the lengths of the regions. The theorem implies that there are only a finite number of regions which occur in a fixed order. The objective function is quadratic in the region lengths and the constraints are linear. However, the size of this quadratic program grows rapidly with n_2 and n_3; the number of variables is approximately factorial (n_2+n_3).

7.6 DUALITY THEORY

As usual we can obtain a weak duality result; the complementary slackness conditions take the form

$$\int_0^T w(t)^{\mathrm{T}} x(t)\, \mathrm{d}t = 0,$$

$$\int_0^T (u(t)^{\mathrm{T}} y(t) + v(t) z(t)\, \mathrm{d}t = 0.$$

Set

$$\hat{u}(t) = \begin{bmatrix} w(t) \\ u(t) \\ v(t) \end{bmatrix}, \qquad t \in [0, T].$$

Then we have the following theorem.

Theorem 7.7 *If $\hat{x}(t)$ is feasible for* SCLP *and $\hat{u}(t)$ is feasible for* SCLP* *with $\hat{x}(t)^{\mathrm{T}}\hat{u}(t) = 0$ almost everywhere in $[0, T]$, then $\hat{x}$ is optimal for* SCLP *and $\hat{u}$ is optimal for* SCLP*.

It is still an open question whether, or in what circumstances, a strong duality result can be proved for problems of the form SCLP. Certainly, the conditions of Theorem 7.3 are not very useful in practice for this type of problem. Moreover, the basic solutions for the separated continuous linear program are degenerate, and so even the construction of the reduced-cost functional is not straightforward.

We finish our treatment of this problem by giving a method which can often serve as an optimality check for SCLP. First we prove a result about the optimal solution to the dual problem, under conditions on b and c which are satisfied for many problems of interest.

Theorem 7.8 *If c is continuous and $b(t)$ is a constant, b, then if there is an optimal solution to* SCLP*, *there is an optimal solution in which v and w are continuous.*

Proof. Suppose that u^*, v^*, w^* are optimal for SCLP*. Write

$$d(t) = c(t) - \int_t^T G^{\mathrm{T}} u^*(s)\, \mathrm{d}s.$$

Then for each t, $v^*(t)$ solves the finite linear program

$$\begin{aligned} &\text{minimize} && b^{\mathrm{T}}v \\ &\text{subject to} && H^{\mathrm{T}}v \geqslant d(t), \\ &&& v \geqslant 0. \end{aligned}$$

Since $d(t)$ is continuous, it is a straightforward result in the theory of parametric linear programming that the optimal solution to this problem can be chosen to vary continuously. Hence $v^*(t)$ and $w^*(t)$ can be chosen to be continuous functions. □

Now we look for a solution to the dual constraints which is complementary to a given basic solution $\hat{x}$, and for which v and w are continuous. Differentiating the dual constraints we obtain, at least almost everywhere,

$$-G^{\mathrm{T}}u(t)+H^{\mathrm{T}}\dot{v}(t)-\dot{w}(t)=\dot{c}(t), \qquad t\in[0,T] \tag{10}$$

together with the boundary condition

$$H^{\mathrm{T}}v(T)-w(T)=c(T). \tag{11}$$

Set

$$\bar{u}(t)=\begin{bmatrix}-\dot{w}(t)\\ u(t)\\ -\dot{v}(t)\end{bmatrix}, \qquad \bar{c}(t)=\begin{bmatrix}\dot{c}(t)\\ 0\\ 0\end{bmatrix}, \qquad t\in[0,T].$$

Then we can rewrite (10) as

$$\bar{c}(t)+K^{T}\begin{bmatrix}u(t)\\ -\dot{v}(t)\end{bmatrix}=\bar{u}(t), \qquad t\in[0,T]. \tag{12}$$

Now suppose that $\bar{u}$ corresponds to a $\hat{u}$ which is complementary slack with $\hat{x}$. If $\hat{x}$ is regional, then within an interval where some component $\hat{x}_i$ is positive the corresponding component $\hat{u}_i$ is zero. Hence $\bar{u}_i$ is also zero. Thus from (12) we have

$$K_{\mathrm{B}}^{\mathrm{T}}(t)\begin{bmatrix}u(t)\\ -\dot{v}(t)\end{bmatrix}+c_B(t)=0,$$

where the subscript B indicates those columns or elements which correspond to the support of $\hat{x}$. If K_B is invertible, then

$$\begin{bmatrix}u(t)\\ -\dot{v}(t)\end{bmatrix}=-(K_B{}^{-1}(t))^{\mathrm{T}}\bar{c}_B(t), \qquad t\in[0,T].$$

Hence, from (12),

$$\bar{u}(t)=\bar{c}(t)-K^{\mathrm{T}}(K_B^{-1}(t))^{\mathrm{T}}\bar{c}_B(t). \tag{13}$$

This determines $\hat{u}$ up to an additive constant in v and w. Usually $\hat{u}$ is determined from (11) and (13), together with the condition that where a component of x or z is non-zero, then the corresponding component of w or v is zero. It may be, however, that $\hat{u}$ is overdetermined; in these circumstances there will be some component of $\hat{u}$ which is non-zero (but with zero derivative) and which corresponds to a non-zero component of $\hat{x}$. If in fact $\hat{u}$ is complementary to $\hat{x}$, then, from Theorem 7.7, $\hat{u} \geqslant 0$ is sufficient for $\hat{x}$ to be optimal.

7.7 NOTES

The continuous linear program (or bottleneck problem) has a large literature associated with it, and we shall briefly review this here. The problem was introduced by Bellman (1957). Since then the published work concerned with this problem has fallen into one of two groups, being primarily concerned either with the duality theory for such problems or with methods for their solution.

In the first group the literature stems from a paper of Tyndall (1965), who gives conditions under which CLP and its dual have the same value when they are both posed in L_∞ spaces. This work was extended by Levinson (1966a) (who gives conditions for the case when B and K are time-dependent matrices) and by Tyndall (1967). Hanson (1968) and Hanson and Mond (1968) give further extensions to the case where the objective function is concave. All this work relies on proofs of the strong duality theorem by considering a sequence of successively finer discretizations. In Grinold (1969a, 1969b) a different approach is taken which is closer to the abstract mathematical programming approach. This is also true of the papers by Murakami and Yamasaki (1969) and Schechter (1972b) where CLP and its dual are posed over other spaces as well as L_∞. These papers both make use of results proved over arbitrary vector spaces (Kretschmer (1961) and Schechter (1972a)). Other papers related to duality theory for CLP are those by Larsen and Polak (1966), Grinold (1970), Farr and Hanson (1974a, 1974b), Underwood (1976), Singh and Farr (1977), Bodo and Hanson (1978), Singh (1978), Levine and Pomerol (1979) and Reiland (1980a, 1980b). Papageorgiou (1982) gives a dual problem posed in a larger space (essentially his dual problem includes measures on $[0, T]$) and is able to obtain a more satisfactory duality theory in this setting. Troltzsch (1980) also considers duality theory for a particular version of CLP.

The second group of papers is concerned with algorithms for the solution of CLP. This work was started by Lehman (1954) and the papers by Drews (1974), Hartberger (1974) and Segers (1974) are along similar lines. The first person to make major progress with the specification of a simplex-like algorithm for CLP was Perold (1978). His work has been continued by Anstreicher (1983). However, the algorithm which they describe is incomplete and exceedingly complex. This reflects the inherent difficulties of the problem. The paper by Sethi *et al.* (1982) relates the optimal dual variables to the adjoint variables appearing when CLP is posed as an optimal control problem with state-variable inequality constraints.

There are a considerable number of papers by Russian authors about CLP. These are mainly concerned with types of solution procedure which are not related to the simplex algorithm. We shall only mention papers which appear in an English translation. One of the first papers in this area is that by Dubovskii *et al.* (1973), who suggest a method based on the maximum principle. A variety of other methods for the solution of CLP are discussed by Charnyi (1975, 1976), Ilyutovich (1976) and Pestryakov (1977). Meerov *et al.* (1980) give a quite complete treatment of certain special classes of infinite-dimensional linear program. Two papers which deal with conditions for optimality of CLP are those by Dyukalov (1976) and Mokhov (1977). Some other related papers are by Dyukalov and Ilyutovich (1977), Chukanov (1976, 1978a, 1978b), Pestryakov (1978, 1980), Charnyi (1978), Boikov and Charnyi (1979), Ilyutovich (1980) and Volkov (1980).

Perold (1981) discusses the extreme points of the set of feasible solutions for CLP and the approach he takes is somewhat similar to that we have followed for SCLP. The material of Sections 7.4 and 7.5 is taken from Anderson, Nash and Perold (1983). An algorithm for SCLP is an attractive possibility. Some preliminary work in this direction can be found in Anderson (1978, 1980).

Problems

7.1 Show that the recession cone for $L(A, y)$ is precisely the set $\{x \geqslant 0 : Ax \leqslant 0\}$.

7.2 Prove directly the result of Theorem 7.7 (i.e. without using Theorem 3.2).

7.3 Show that, if conditions (A) and (B) are modified appropriately, Theorem 7.3 will hold when the matrix B is given as a function of time $B(t)$.

7.4 Formulate the general continuous linear program as a linear optimal control problem in the same manner as was done for SCLP in Section 7.3.

7.5 Prove that (as stated in the proof of Theorem 7.8) for a finite linear program LP(b), taken to depend on the right-hand side vector b, if b is given as a continuous function of a variable s, then we can find a continuous function $x^*(s)$, such that $x^*(s)$ is optimal for LP($b(s)$).

7.6 Find the space $D(\hat{x})$, when $\hat{x}$ is a basic solution for SCLP, and hence show that $\hat{x}$ is in general degenerate.

8

Other Infinite Linear Programs

8.1 INTRODUCTION

In the last four chapters, we have looked in some detail at four specific classes of infinite linear programs, and seen how the theory developed in Chapters 2 and 3 applies in those particular cases. Many other infinite linear programs can be posed, and in this final chapter we shall briefly survey some of them. The theory of these problems has been rather less fully worked out, and our discussion of them is to an extent a survey of some of the avenues of research which invite attention at this stage in the development of the theory of infinite linear programming.

We shall look first at a linear programming formulation which is of considerable generality, capable, with an appropriate choice of the underlying spaces, of encompassing several of the examples treated earlier. One example is closely connected with the problem of determining the electrostatic capacity of a conducting body, and for this reason problems of this type have been called *capacity* problems. We then examine two further extensions of the network-flow problem. One of these has similarities with the maximum-flow problem discussed in Chapter 6; this time, however, we consider the case when there is a cost associated with flows in each arc, and the objective is to minimize the total cost of the flow, subject to certain demand and supply constraints at the nodes. The second is an extension of the maximum-flow problem obtained by allowing the nodes of the network to increase in number, so that the network becomes a continuum. Thus we seek to maximize flow in some domain subject to an upper bound on flow density at each point, and a mass-conservation constraint expressed by the condition that the flow be divergence free in the interior of the domain.

Our final application is to a problem arising in the optimal design of structures. In the study of structures by means of finite-element methods, there are several examples of physical laws expressible as complementary variational principles in the form of a linear programming duality. The treatment of these principles by the methods of infinite-dimensional linear programming remains an open research area of considerable theoretical interest, and potential practical importance. The area is fraught with difficulties, however, and as yet little progress has been made. The whole theory of infinite-dimensional LP has itself emerged as an important topic only recently; it seems appropriate, therefore, to end by touching on a research area which is so far almost totally unexplored.

8.2 THE CAPACITY PROBLEM

In electrostatics, the capacity of a conducting body is defined as the ratio of the total charge on the body to the electrical potential of the body; this is well-defined because the potential in a (perfect) conductor is constant. The self-energy of the charge on the body is eV, where e is the total charge and V is the potential. The distribution of charge is determined by the shape of the body (although, by another theorem of electrostatics, it is all on the surface), and will naturally be such as to minimize this energy. Since the charge is constant, this is equivalent to minimizing V for a given value of e. Equally, we could derive the charge distribution by maximizing the total charge subject to an upper bound on the potential. In this latter form, the problem is an infinite-dimensional linear program:

$$\text{ECP:}\quad \text{maximize} \quad \int_\Gamma \varepsilon(\xi)\,\mathrm{d}\Gamma(\xi)$$

$$\text{subject to} \quad \int_\Gamma \frac{\varepsilon(\xi)}{|(\xi - x)|}\,\mathrm{d}\Gamma(\xi) \leqslant 1, \qquad \text{for all } x \in \Gamma,$$

where Γ is the surface of the body. We can formulate a dual to this problem:

$$\text{ECP*:}\quad \text{minimize} \quad \int_\Gamma p(\xi)\,\mathrm{d}\Gamma(\xi)$$

$$\text{subject to} \quad \int_\Gamma \frac{p(\xi)}{|(x - \xi)|}\,\mathrm{d}\Gamma(\xi) \geqslant 1, \qquad \text{for all } x \in \Gamma.$$

This problem exemplifies a natural generalization of the finite linear program, in which the matrix A is replaced by the kernel of an integral operator, and the decision vector is replaced by some function or distribution over a given set. The most striking feature of the pair of problems posed above is the similarity of form of the primal and dual. This feature is retained in a general class of problems of this form called *capacity problems*. Suppose U and V are two sets, and M is a real-valued function on $U \times V$, continuous in some topology, which we need not specify. The capacity problem is posed in the space of regular Borel measures on U:

$$\text{CAP:}\quad \text{maximize} \quad \int_U b(\xi)\,\mathrm{d}\nu(\xi)$$

$$\text{subject to} \quad \int_U M(\xi, \zeta)\,\mathrm{d}\nu(\xi) \leqslant c(\zeta), \qquad \text{for all } \zeta \in V,$$

$$\nu \geqslant 0.$$

We pose a dual problem for CAP in the space of regular Borel measures on V.

$$\text{CAP}^*: \quad \text{minimize} \quad \int_V c(\zeta)\,\mathrm{d}\mu(\zeta)$$

$$\text{subject to} \quad \int_V M(\xi,\zeta)\,\mathrm{d}\mu(\zeta) \geqslant b(\xi), \qquad \text{for all } \xi \in U,$$

$$\mu \geqslant 0.$$

These two problems are nearly identical, so we can choose to study either. We will study the dual, to conform with our previous convention of studying minimization problems.

For concreteness and ease of exposition, we will consider an example in which both U and V are the unit interval of the real line. We will assume that M and b are analytic functions, and let

$$z(x) = \int_0^1 M(x,y)\,\mathrm{d}\mu(y) - b(x).$$

It can be shown that the structure of the extreme points of the feasible set is in this case similar to that for the continuous semi-infinite program described in Section 4.4. Under the assumptions we have made here, it is clear that the optimal value can be approximated arbitrarily closely by a basic solution whose support is concentrated on a finite set of points. In general for such a solution, z has only a finite number of zeros. Let $\{y_1, y_2, \ldots, y_n\}$ be the support of μ, and let $x_1, x_2, \ldots, x_m$ be the zeros of z. Let d_i be the order of the zero at x_i, so that

$$z^{(k)}(x_i) = 0, \qquad k = 0,1,2,\ldots,d_i,$$

with $z^{(k)}(x_i)$ not equal to 0 for $k = d_i + 1$. Note that the first non-zero derivative must be positive, if x_i is in the interior of $U = [0,1]$, since a zero of z is a minimum if μ is feasible.

If we work entirely with analytic functions, then the space $B(z)$ consists of functions with zeros of order at least d_i at each of the points x_i, $i = 1,2,\ldots,m$ (see Lemma 4.9). The space $B((\mu, z))$ is the product of the space of measures whose support is contained in $\operatorname{supp}(\mu)$ and $B(z)$. The intersection of this product with the null space of the constraint map consists of measures μ' in $B(\mu)$ such that

$$\int_0^1 M(x,y)\,\mathrm{d}\mu'(y) = z'(x) \in B(z).$$

If μ'_j is the mass of μ' at y_j, $j = 1,2,\ldots,m$, then this condition may be re-written

$$\sum_{j=1}^{m} M(x, y_j)\,\mu'_j = z'(x) \in B(z).$$

This in turn implies the following conditions:

$$\sum_{j=1}^{n} \frac{\partial^k}{\partial x^k} M(x_i, y_j)\,\mu'_j = 0, \qquad k = 0,1,2,\ldots,d_i, \quad i = 1,2,\cdots,m.$$

If μ is basic, then these relations must imply that $\mu' = 0$. Denote by $\tilde{M}$ the matrix whose columns are the vectors

$$\left[\frac{\partial^k}{\partial x^k} M(x_i, y_1), \frac{\partial^k}{\partial x^k} M(x_i, y_2), \ldots, \frac{\partial^k}{\partial x^k} M(x_i, y_n)\right]^{\mathrm{T}}, \qquad k = 0,1,2,\ldots, d_i, \quad i = 1,2,\ldots, m.$$

We can now state a theorem which corresponds closely to Theorem 4.10.

Theorem 8.1 *If μ is feasible for CAP*, and supp(μ) is a finite set of points, then μ is basic if and only if the rows of $\tilde{M}$ are linearly independent.*

By way of illustration, consider the following example. Take $U = V = [-1,1]$, let b and c be the constant functions 1, and let

$$M(x, y) = ((x-y)^2 - 2)^2 .$$

Then the problem CAP* becomes

$$\text{minimize} \quad \int_{-1}^{1} \mathrm{d}\mu(x)$$

$$\text{subject to} \quad \int_{-1}^{1} ((x-y)^2 - 2)^2 \, \mathrm{d}\mu(x) \geqslant 1, \qquad -1 \leqslant y \leqslant 1,$$

$$\mu \geqslant 0.$$

Consider the feasible solution given by a three-point measure $\mu(-1) = \mu(1) = \frac{1}{9}$, $\mu(0) = \frac{2}{9}$. A calculation shows that the constraint is satisfied with equality at the two points $y_1 = 1/\sqrt{2}$ and $y_2 = 1/\sqrt{2}$. Thus $m = 2$, and $d(1) = d(2) = 1$. Hence

$$\tilde{M} = \begin{bmatrix} M(-1, y_1) & M_x(-1, y_1) & M(-1, y_2) & M_x(-1, y_2) \\ M(0, y_1) & M_x(0, y_1) & M(0, y_2) & M_x(0, y_2) \\ M(1, y_1) & M_x(1, y_1) & M(1, y_2) & M_x(1, y_2) \end{bmatrix}.$$

It is straightforward to check that the rows of this matrix are linearly independent, and so μ is a basic solution. Normally we should expect $\tilde{M}$ to be square, but in this case μ is degenerate. We have been able to satisfy four conditions by the choice of only three values of μ. Actually this μ is optimal as the reader may check by considering a solution to CAP given by $\nu(y_1) = \nu(y_2) = \frac{1}{11}$.

It is usually the case that an optimal solution to CAP has its support concentrated on a finite set of points. The opposite is true for the electrostatic problem, ECP, where the optimal distribution of charge is spread out over the surface of the body; treated as a measure on Γ, the charge is absolutely continuous. The reason for this is that in ECP the function corresponding to $M(x, y)$ is not continuous, being unbounded at $x = y$.

8.2.1 Notes

The electrostatic capacity problem and other related physical problems have been studied intensively for many years. From a mathematical standpoint, these problems have given rise to the great apparatus of modern potential theory, associated with the names of, for example, Riesz (1938) and Choquet (1953). This is a branch of mathematics within which the capacity of a set is frequently defined using the supremum of some quantity over a set of feasible measures, but which does not have any concern with questions of optimization. The recognition of the capacity problem as a general linear programming problem, and its study as such, was first made by some Japanese mathematicians, Yamasaki (1966) and Ohtsuka (1966a, 1966b). The result we have given characterizing the extreme points of the feasible sets is due to Wu (1985). Wu also discusses a solution method for CAP based around the observation that if the points of support for a feasible solution are fixed, then the optimal values can be found from a semi-infinite linear program.

8.3 THE CONTINUOUS NETWORK PROGRAM

In Chapter 6 we dealt with the maximal flow problem in a dynamic network with storage. In this section we again consider a dynamic network with storage at the nodes; but we shall look at a more general problem having a cost on the flow in each arc of the network. Specifically we consider a network Ω with p nodes and q directed arcs; the structure of Ω is described by the node–arc incidence matrix A. This is a $p \times q$ matrix with $A_{ij} = 1$ if the jth arc comes from node i, $A_{ij} = -1$ if the jth arc goes into node i, and $A_{ij} = 0$ otherwise. We define the continuous network program as follows

$$\begin{aligned} \text{CNP:} \quad \text{minimize} \quad & \int_0^{T} c(t)^{\mathrm{T}} x(t)\,\mathrm{d}t \\ \text{subject to} \quad & r(t) \leqslant \int_0^t Ax(\tau)\,\mathrm{d}\tau \leqslant s(t), \qquad t \in [0, T], \\ & 0 \leqslant x(t) \leqslant u(t), \qquad t \in [0, T]. \end{aligned}$$

Here $x(t) \in R^q$ and $u(t) \in R^q$ give the arc flows and arc capacities at time t. Let

$$y_i(t) = \int_0^t (Ax(\tau))_i\,\mathrm{d}\tau, \qquad i = 1, 2, \ldots, n, \quad t \in [0, T].$$

Then $y_i(t)$ gives the total flow into node i up to time t, and is constrained to lie between $r_i(t)$ and $s_i(t)$. Thus the vector $s(t) - r(t)$ gives the node storage capacities at time t. $r_i(t)$ can be interpreted as the total demand at node i up to time t, so that a source node will have $r_i(t) < 0$. The initial amounts stored at the nodes at time 0 are given by $-r(0)$; Clearly $r(0) \leqslant 0$ is a necessary condition for the problem to be consistent. We shall assume that the functions r, s and u are all continuous.

The formulation of CNP is very flexible, and a variety of dynamic network problems can be put into this form. For example the maximal flow problem of Chapter 6 can be reformulated as an example of CNP by adding a return arc, going from the sink to the source, having infinite capacity and a negative cost, while all the other arcs have zero cost. Other examples of dynamic network problems may have costs on the amount stored at the nodes of the network. Such problems can be transformed into CNP using an integration by parts.

The simplex algorithm specialized to a network offers an extremely effective way of solving finite-dimensional network programs. The reason for this lies in the form which the basic solutions take for the network program. At a basic feasible solution the arcs containing flow which is both non-zero and sub-capacity form a set of trees. In the non-degenerate case they form a single spanning tree for the network. As we shall see below a similar characterization is possible for CNP. If x is feasible for CNP, we say that an arc k or a node i is *free* at time t if the corresponding variable is strictly between its bounds, i.e. if $0 < x_k(t) < u_k(t)$ or $r_i(t) < y_i(t) < s_i(t)$ respectively.

It is easy to see that CNP is an example of the separated continuous linear program introduced in Chapter 7. This enables us to give a characterization of the basic solutions for CNP, when it is posed over $L^n_\infty[0, T]$, using Theorem 7.4.

Theorem 8.2 *A feasible solution x for* CNP *is basic if and only if, for almost all $t \in [0, T]$, the free arcs form a set of trees in each of which there is at most one free node.*

Proof. Let K be the $(2p+q)\times(2p+2q)$ matrix defined by

$$K = \begin{bmatrix} -A & I & 0 & 0 \\ A & 0 & I & 0 \\ I & 0 & 0 & I \end{bmatrix}.$$

From Theorem 7.4, a feasible solution x is basic if and only if the columns of K corresponding to non-zero variables are linearly independent for almost all $t \in [0, T]$. The four groups of columns of K correspond to $x(t)$ itself and the three sets of slack variables given by $y(t) - r(t)$, $s(t) - y(t)$ and $u(t) - x(t)$ respectively.

Suppose first of all that there is a cycle of free arcs at time t. Traversing the cycle in one or other direction defines a vector $v \in R^q$, having $v_k = 1$ for arcs k in the cycle and directed in the same way as the traversal, $v_k = -1$ for arcs k in the cycle and not directed in the same way as the traversal, and $v_k = 0$ for arcs not in the cycle. Thus $Av = 0$, and if we define $w \in R^{2p+2q}$ by $w^{\mathrm{T}} = [v^{\mathrm{T}}, 0^{\mathrm{T}}, 0^{\mathrm{T}}, -v^{\mathrm{T}}]$, where 0 is taken as a vector in R^p, then it is clear that $Kw = 0$. However, the support of w is contained within the support of $x(t)$ and the support of $u(t) - x(t)$, hence the appropriate columns of K are not linearly independent.

On the other hand suppose that, at time t, there is a chain of free arcs between two free nodes, i and j, say. Define $v \in R^q$ by setting $v_k = 1$ if arc k is in the chain and directed from i to j, $v_k = -1$ if arc k is in the chain and directed from j to i, and $v_k = 0$ otherwise. Then $Av = e_i - e_j$, where e_i denotes the ith unit vector in R^p. Letting $w^{\mathrm{T}} = [v^{\mathrm{T}}, (e_i - e_j)^{\mathrm{T}}, (e_j - e_i)^{\mathrm{T}}, -v^{\mathrm{T}}]$ yields $Kw = 0$, and so in this case also the appropriate columns of K are not linearly independent. This establishes the theorem in one direction.

Now suppose that $Kw = 0$ with w non-zero, and let $w^T = [w_1^T, w_2^T, w_3^T, w_4^T]$ with $w_1, w_4 \in R^q$ and $w_2, w_3 \in R^p$. Thus $w_2 = Aw_1$, $w_3 = -Aw_1$ and $w_4 = -w_1$. We shall establish that amongst the arcs k with $(w_1)_k$ and $(w_4)_k$ not equal to zero there is either a cycle (undirected), or a chain joining nodes i and j with $(w_2)_i$, $(w_2)_j$, $(w_3)_i$ and $(w_3)_j$ all non-zero. Choose any arc k with $(w_1)_k$ non-zero. Then observe that, for the nodes at each end of the arc, since $w_2 = Aw_1$, either there is another incident arc with w_1 non-zero or w_2 is non-zero for that node. This establishes the properties we want for w_1 and w_2. This result for w_3 and w_4 follows immediately.

If the columns of K corresponding to non-zero variables are not linearly independent, then we can find a w with $Kw = 0$ and with w_1 and w_4 non-zero only for free arcs, w_2 and w_3 non-zero only for free nodes. Thus we have established the theorem in the other direction. □

8.3.1 Notes

Problems related to CNP have been studied by a number of authors. One example is the dynamic routing problem which seeks a set of flows to clear an accumulation of stock at each storage node of a network, in such a way that the delay cost is minimized. This problem, which has been studied by for example Segall (1977) and Hajek and Ogier (1984), can be easily converted into an example of CNP.

The characterization of basic solutions given above can be derived directly, without the use of the theory of separated continuous linear programs, and this is the approach taken by Philpott (1985). It is natural to look for an extension of the network simplex method to CNP. The essential idea will be to define a potential function $\pi_i(t)$, for each node i, using the following rules:

(a) $d\pi_i(t)/dt = 0$ if node i is free at time t;
(b) $\pi_i(t) - \pi_j(t) = c_k(t)$ if the kth arc goes from node i to node j and is free at time t.

If a solution is basic, then the application of these rules cannot produce a contradiction. It is not hard to show that if π satisfies (a) and (b), then the current solution is optimal if (i) $y_i(t) = r_i(t)$ ($s_i(t)$) whenever π_i is increasing (decreasing) at t, and (ii) if the kth arc goes from node i to node j, then $x_k(t) = 0$ ($u_k(t)$) whenever $c_k(t) - \pi_i(t) + \pi_j(t)$ is strictly greater than (less than) 0. A simplex-like algorithm for CNP is obtained by finding a construction to give an improved basic feasible solution if either (i) or (ii) fails to hold. The details of this algorithm are worked out in Anderson and Philpott (1986).

8.4 THE SPACE-CONTINUOUS MAXIMUM-FLOW PROBLEM

In Chapter 6 we saw how the maximum-flow minimum-cut theorem of Ford and Fulkerson could be generalized to a time-varying network with storage at the nodes. In this section we shall look briefly at another appealing infinite-dimensional generalization of the Ford–Fulkerson theory. We consider allowing the nodes of the network to become infinite in number, so that the network becomes a continuum. In fact we seek a flow defined over a region of the plane, subject to certain capacity constraints which operate at each point of the region, and ask for this flow to maximize the total flow between two given subsets of the boundary of the region, called the source and sink, respectively.

Let Ω be a compact connected region in R^2 with non-empty interior and piecewise smooth boundary $\partial\Omega$, and let the source Γ_p and the sink Γ_q be disjoint subsets of $\partial\Omega$. The flow in Ω is given by a vector field $\boldsymbol{\sigma} = (\sigma_1, \sigma_2)$, with the norm of $\boldsymbol{\sigma}$ at $x \in \Omega$ defined by $\|\boldsymbol{\sigma}(x)\| = \sqrt{(\sigma_1(x)^2 + \sigma_2(x)^2)}$. Let $\Gamma_r = \partial\Omega - \Gamma_p - \Gamma_q$, in other words that part of the boundary which is neither source nor sink. Denote by $\mathbf{n}$ the unit outward normal to $\partial\Omega$, where this is defined.

The space-continuous maximum-flow problem is formulated as follows:

$$\begin{aligned}
\text{SCNP:}\quad &\text{maximize} && \int_{\Gamma_q} \boldsymbol{\sigma}\cdot\mathbf{n}\,\mathrm{d}s \\
&\text{subject to} && \nabla\cdot\boldsymbol{\sigma} = 0 \text{ in } \Omega, \\
& && \boldsymbol{\sigma}\cdot\mathbf{n} = 0 \text{ on } \Gamma_r, \\
& && \|\boldsymbol{\sigma}(x)\| \leqslant c(x), \qquad \text{for } x \in \Omega.
\end{aligned} \tag{1}$$

In order for $\boldsymbol{\sigma}$ to satisfy (1), both σ_1 and σ_2 must have continuous derivatives. This is rather a stringent condition, and there is no *a priori* reason to suppose that an optimal solution would have this property. For this problem it is more natural to work with flows which are piecewise well-behaved. Because we are working over a region of the plane, some rather technical definitions are necessary to achieve this end.

First define a *piecewise* C^2 curve to be a piecewise smooth curve with bounded curvature at the points where its derivative is continuous. Suppose that Ω is divided up into a finite number of regions $R_1, R_2, \ldots, R_M$, each having a piecewise C^2 boundary, and such that σ_1 and σ_2 have continuous and bounded first partial derivatives within each R_j; then $\boldsymbol{\sigma}$ is called a *regular* flow. We now repose SCNP in the class of regular flows, and substitute for (1) the constraint,

$$\int_\gamma \boldsymbol{\sigma}\cdot\mathbf{n}_\gamma\,\mathrm{d}s = 0,$$

for all piecewise smooth closed curves γ in Ω with outward normal $\mathbf{n}_\gamma$. This is just a way of using the divergence theorem to get around the fact that $\boldsymbol{\sigma}$ does not have derivatives everywhere.

We shall outline the steps necessary to deduce a version of the maximum-flow minimum-cut theorem, for SCNP posed in this way. The proof is analogous to that of the classical result, but considerable care is necessary in the constructions that are used. For the moment we assume that Γ_p and Γ_q are open in $\partial\Omega$ and have a finite number of components. The capacity function c is assumed to be non-negative and continuous with piecewise continuous and bounded first partial derivatives in Ω.

In the classical case, cuts are defined by partitioning the nodes of the network into two complementary sets, one containing the source and the other containing the sink. The value of the cut is defined as the sum of the capacities of arcs joining nodes in the set containing the source to nodes in its complement. The optimal flow is constructed by a labelling procedure; on termination this defines a cut with minimum value. The extension of these constructions to the space-continuous case requires some care. The rules which define the labelling procedure must be sufficiently powerful to ensure that a feasible increase in flow can be made from the source Γ_p to any labelled point in Ω. Furthermore, the definition of a cut must be general enough for the set of points which are reachable upon termination of the labelling procedure to constitute a well-defined cut. The formal definitions are made as follows.

Any set in $\Omega - \Gamma_q$ which contains Γ_p is called a *cut* in Ω. If a cut C has a non-empty interior D in R^2, we define sets A and B by

$$A = \partial D - \Gamma_p - \Gamma_r, \qquad B = \Gamma_p - \partial D,$$

where ∂D is the boundary of D. The sets A and B are illustrated in Figure 8.1. When $D = \varnothing$, we set $A = \varnothing$, and $B = \Gamma_p$. The value $V(C)$ of the cut is defined only when every component of $A \cup B$ is a rectifiable Jordan arc, and $A \cup B$ has

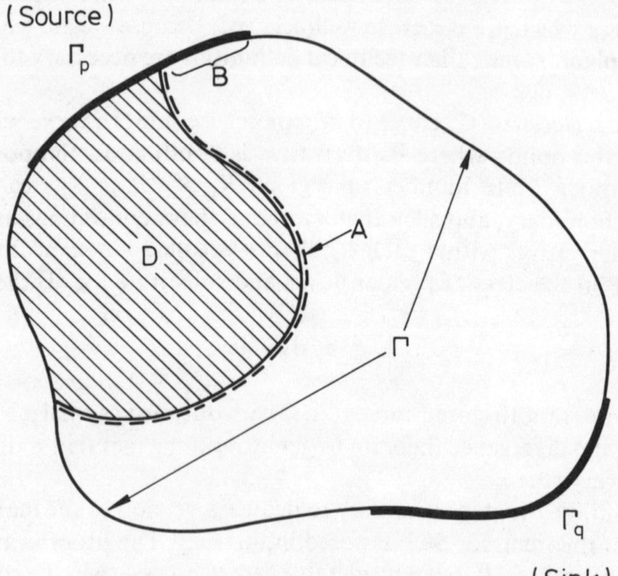

Figure 8.1 A cut for the space-continuous maximum-flow problem

finite length. In this case

$$V(C) = \int_A c\,\mathrm{d}s + \int_B c\,\mathrm{d}s.$$

For any feasible flow σ in Ω, we define its value, $v(\sigma)$ by

$$v(\sigma) = \int_{\Gamma_q} \sigma \cdot \mathbf{n}\,\mathrm{d}s.$$

The following lemma is a simple extension of the corresponding result in the classical theory.

Lemma 8.3 *If σ is a regular feasible flow in Ω with value $v(\sigma)$, and C is a cut in Ω with value $V(C)$, then $v(\sigma) \leqslant V(C)$.*

Now suppose that σ is a regular feasible flow in Ω. A point $x \in \Omega$ is called *reachable* if there is a piecewise C^2 simple arc r, smooth within each R_j, which satisfies the following conditions.

(a) $r(0) \in \Gamma_p$, $r(s^*) = x$ for some s^*;
(b) r meets each ∂R_j at only a finite number of points, at each of which ∂R_j makes a non-zero angle with r, ∂R_j is smooth and r crosses the region boundary;
(c) there is some $\varepsilon > 0$ such that

$$\|\varepsilon t(s) + \sigma(r(s))\| \leqslant c(r(s)),$$

for every value of s for which $t(s)$, the unit tangent vector to $r(s)$, is defined.

The second condition in this definition is purely technical, and of no great importance. The third condition is necessary to ensure that the current flow σ can be increased by an amount ε in the tangent direction along the arc r. It turns out that in these circumstances we can increase the flow in a small corridor around r and remain feasible. This if σ is a regular feasible flow for Ω with value $v(\sigma)$, and some $x \in \Gamma_q$ is reachable, then there is a regular feasible flow σ' in Ω with a value $v(\sigma') > v(\sigma)$.

We now address the problem of constructing a flow σ and a cut C having equal values. Suppose that, for some regular feasible flow σ with value $v(\sigma)$, there is no reachable point in Γ_q. It follows that the set $C = \{x\text{: } x \text{ is reachable}\}$ is a cut in Ω. We wish to show that C has value $v(\sigma)$, but in order to define $V(C)$ we must first demonstrate that ∂D is sufficiently well behaved. To do this it is necessary to make some further assumption about the function c, for example that c is bounded away from zero on Ω. In this case it can be shown that the components of ∂D are Jordan curves and are piecewise smooth in the interior of each region. Once this is established, the result we require, that $V(C) = v(\sigma)$, can be proved fairly readily. Then the following maximum-flow minimum-cut theorem is immediate.

Theorem 8.4 *Suppose that* SCNP *has a regular optimal solution and c is bounded* *away from zero. Then the maximum value of the flow through* Ω *is equal to the minimum value of all cuts in* Ω.

8.4.1 Notes

There is a wide variety of space-continuous versions of the network-flow problem. In general these problems are concerned with finding the flow of some commodity through a region of n-dimensional Euclidean space. For most applications the region Ω will be some compact subset of R^2 or R^3, with a sufficiently well-behaved boundary $\partial\Omega$. Problems of this general type have been considered by Hu (1969), Gomory *et al.* (1973), Iri (1979) and Strang (1983).

To ensure the solvability of SCNP, it is likely that a more abstract formulation is required. Temam and Strang (1980) have posed a version of SCNP in $L_2(\Omega)$; in this case the divergence operator is defined in a distributional sense and solvability follows from a weak* compactness argument. Jacobs (1979) and Jacobs and Seiffert (1983) have established a similar existence result for a measure-theoretic formulation of the problem where the flows are measures of Lipschitz paths joining the source to the sink.

The details of the theory that we have sketched in this section can be found in Philpott (1982). An alternative approach to the maximum-flow minimum-cut theorem involves duality theory, rather than the labelling scheme we have described. This is the approach that is taken by Strang (1983), who establishes a maximum-flow inflimum-cut theorem. He considers a problem in which sources and sinks are distributed over both the interior and boundary of Ω, and he chooses to maximize a scale factor λ by which the flows in and out of the region are multiplied. Thus he considers constraints of the form $\boldsymbol{\sigma}\,.\,\mathbf{n} = \lambda f$ on $\partial\Omega$, and $\nabla\,.\,\boldsymbol{\sigma} = \lambda F$ within Ω, where f and F are given functions.

8.5 OPTIMAL DESIGN OF STRUCTURES

The design of a structure which is the most economical possible, given certain constraints, is a problem of perennial interest to engineers. Here we shall look at one such problem, that of the optimal design of a truss or braced framework. We begin by formulating this design problem as a finite linear program, and then consider extending our formulation to a framework which may have an infinite number of members of infinitesimal length. At first sight this seems a rather impractical problem to consider. It is an example of a limiting solution to an optimization problem, which cannot actually be attained. If, for example, we wish to remove the maximum amount of material from a metal component, while still retaining some given strength, then it is best to use some type of honeycomb structure; and the finer the honeycomb, the more material we shall be able to remove. Thus the optimal solution represents an ideal which will be approximated to a greater or lesser degree in practice.

Consider first a framework which transmits a given set of forces or loads to a foundation (e.g. a bridge). If the framework is statically determinate, the forces in each member are given by the solution to the equilibrium equations at each node or joint. If the joints are numbered $1, 2, \ldots, N$, and the member joining i to j is in the direction of the unit vector u_{ij}, then the equilibrium equations are

$$\sum_{j=1}^{N} \sigma_{ij} u_{ij} = f_i, \tag{2}$$

for each joint i that is not a foundation point. Here f_i is the vector load applied at node i and σ_{ij} is a scalar giving the force in the member joining node i to node j (this is positive if the member is in tension, and negative if in compression). This formulation ignores the self-weight of the members, as we shall continue to do throughout our discussion.

We may suppose that the material of the members can sustain a certain maximum magnitude of tension per unit cross-section and a certain maximum magnitude of compression per unit cross-section. For simplicity we take these as the same number, so that the minimal cross-sectional area of a member is just proportional to $|\sigma_{ij}|$. Thus the total volume of material in the structure is proportional to the sum over all the members (i,j) of $r_{ij}|\sigma_{ij}|$, where r_{ij} is the magnitude of the distance from node i to node j.

The optimal design problem is to choose the positions of the joints and the layout of the members so as to minimize this total volume of material. If we take the joint positions as fixed and consider varying the members which are included, then we arrive at a problem in which the σ_{ij} are the variables. In fact we have the linear program:

$$\begin{aligned}
\text{LPS:}\quad &\text{minimize} && \sum_{j=1}^{N}\sum_{i=1}^{N} r_{ij}\tau_{ij} \\
&\text{subject to} && \sum_{j=1}^{N} \sigma_{ij} u_{ij} = f_i, \qquad i \in Q, \\
& && \sigma_{ij} - \sigma_{ji} = 0, \qquad i \neq j, \\
& && \tau_{ij} - \sigma_{ij} \geqslant 0, \\
& && \tau_{ij} + \sigma_{ij} \geqslant 0, \\
& && \tau_{ij} \geqslant 0, \qquad i,j = 1,2,\ldots,N.
\end{aligned}$$

Here Q denotes the set of points which are not anchored, and τ_{ij} will be the modulus of σ_{ij} at the optimal solution. As before, u_{ij} and f_i are vectors, the other quantities are scalars. Note that the objective function in LPS counts every member twice.

A framework which has more members than necessary does not have the force in each member determined uniquely by (2). In this case an analysis of the structure has to consider extensions of the members (and uses Hooke's law). If the forces in the members are determined exactly by (2), then the structure is called *statically determinate*. This concept is related to the extreme points structure of the linear program LPS by the following easy result.

Theorem 8.5 *With respect to a given set of forces f_i, the non-degenerate basic feasible solutions for LPS are exactly those in which the members (i,j) having non-zero σ_{ij} from a statically determinate network.*

The dual problem for LPS can be readily written down; some straightforward manipulation gives the following

$$\begin{aligned} \text{LPS*:}\quad & \text{maximize} && \sum_{i\in Q} y_i^{\mathrm{T}} f_i \\ & \text{subject to} && y_i = 0, \quad i \notin Q, \\ & && -2r_{ij} \leqslant (y_i - y_j)^{\mathrm{T}} u_{ij} \leqslant 2r_{ij}. \end{aligned}$$

The 2s in the second constraint correspond to the factor of 2 in the primal objective function.

Now suppose that the joints of the framework can be positioned wherever we like, and that there may be as many joints as we wish. This gives the potential for a truss-like continuum or web. In this situation the primal problem must be formulated in terms of the stress tensor. At this point it is convenient to restrict ourselves to planar problems, for which the stress tensor is represented by the symmetric matrix

$$S = \begin{bmatrix} \sigma_x & \tau_{xy} \\ \tau_{xy} & \sigma_y \end{bmatrix}.$$

Suppose that the structure is confined to a region, Ω, of the plane. Within this region the equilibrium condition becomes

$$(\partial/\partial x)\sigma_x + (\partial/\partial_y)\tau_{xy} = 0, \tag{3}$$

$$(\partial/\partial x)\tau_{xy} + (\partial/\partial y)\sigma_y = 0. \tag{4}$$

We replace the forces f_i, operating at a finite number of joints, with a force distributed along the boundary Γ of Ω. If the unit normal to Γ is given by $n = (n_1, n_2)$ and there is a force $f = (f_1, f_2)$ acting on Γ, then we need the conditions

$$\sigma_x n_1 + \tau_{xy} n_2 = f_1, \tag{5}$$

$$\tau_{xy} n_1 + \sigma_y n_2 = f_2, \tag{6}$$

on Γ.

Instead of the cross-sectional area of a member being proportional to $|\sigma_{ij}|$, it is optimal to place members at every point in the direction of the principal stresses (the eigenvectors of S) at that point, and these members should have cross-section proportional to the modulus of the respective eigenvalues. Thus we obtain the problem of minimizing

$$\iint_{\Omega} (|\lambda_1(S)| + |\lambda_2(S)|)\,\mathrm{d}x\,\mathrm{d}y,$$

subject to (3)–(6), where $\lambda_1(S)$ and $\lambda_2(S)$ are the eigenvalues of S. It is possible, with some effort, to pose this problem as a linear program, but the procedure is neither elegant nor edifying. Interestingly enough the dual problem is much more accessible.

In the limit, the continuous form of LPS* has y defined as a vector field over Ω. The objective function becomes

$$\int_{\Gamma} y^{\mathrm{T}} n \, \mathrm{d}s.$$

We need to consider what happens to the constraint

$$|(y_i - y_j)^{\mathrm{T}} u_{ij} / r_{ij}| \leqslant 2,$$

in the limit, as the number of nodes becomes infinite. It can be rewritten

$$|(\Delta(y^{\mathrm{T}} u)/|\Delta x|)| \leqslant 2,$$

where u is a unit vector in the direction Δx. Define a map D from vector fields over Ω to scalar fields over $\Omega \times \Pi$, where Π is the unit circle, by

$$(Dy)(x, u) = \lim_{\delta \to 0} \{(1/\delta)(y(x + \delta u) - y(x))^{\mathrm{T}} u\}.$$

Thus Dy is a type of directional derivative of y. Observe that D is a linear map. Hence we obtain the following linear program, as a continuous version of LPS*:

$$\begin{aligned} \text{CLPS*:} \quad & \text{maximize} \quad && \int_{\Gamma} y^{\mathrm{T}} n \, \mathrm{d}s \\ & \text{subject to} && y = 0 \text{ on the foundation,} \\ & && -2 \leqslant (Dy)(x, u) \leqslant 2, \qquad x \in \Omega, \quad u \in \Pi. \qquad (7) \end{aligned}$$

If we write $E(x)$ for the symmetric matrix whose (i, j)th element is

$$[(\partial y_i / \partial x_j) + (\partial y_j / \partial x_i)]/2$$

evaluated at x, then it is not hard to see that $(Dy)(x, u) = u^{\mathrm{T}} E(x) u$. This attains its extreme values over $u \in \Pi$ at the eigenvectors of $E(x)$, where its value is given by the corresponding eigenvalue. Thus, writing $e_1(x)$ and $e_2(x)$ for the eigenvalues of $E(x)$, we can rewrite the constraints (7) as

$$|e_1(x)| \leqslant 2, \quad |e_2(x)| \leqslant 2, \qquad x \in \Omega.$$

Though this procedure greatly simplifies the form of the problem, it is at the cost of the linear structure. It is an open question whether the linear programming formulation of the dual problem can be used to find effective solution procedures for this problem.

8.5.1 Notes

For a treatment of the whole subject of optimal structural design, the reader is referred to the books by Hemp (1973) and Prager (1974). The material of this section, relating to the optimal design of braced frameworks, goes back to a paper of Michell (1904). It is possible, from the formulation of the dual problem that we have given, to derive quite detailed conclusions about the form of an optimal solution to the problem. The infinitesimal members of the structure must lie along curves in a type of orthogonal net, called a Hencky–Prandtl net. The optimal structure that arises is called a *Michell structure*. A lucid treatment of this area can be found in Whittle (1971). Strang and Kohn (1983) discuss some related problems.

Hencky–Prandtl nets also occur in the theory of plasticity, where the shear lines in perfectly plastic solids have this property. Strang and Kohn show that a link can be made between these two theories by considering an optimal design problem in which there is a limit on the strength of individual members (or equivalently, a limit on their cross-section).

Some related work is concerned with equilibrium problems in continuum mechanics which have biextremal formulations; see, for example, Christiansen (1980), Temam (1983) and Kortanek (1985). Kortanek shows how generalized finite element methods for such problems are related to the generalized finite sequence spaces which occur in the duals to certain semi-infinite linear programs.

Bibliography

ABEL, U., DEUBER, W., and TAAKE, G. (1979). A note on time dependent networks, *Operations Research Verfahren*, **36**.

ALLEN, G. (1977). Variational inequalities, complementarity problems and duality theorems, *J. Math. Anal. Appl.*, **58**, 1–10.

ANDERSON, E. J. (1978). *A Continuous Model for Job-shop Scheduling*, Unpublished Ph. D. thesis, University of Cambridge.

ANDERSON, E. J. (1980). Basic solutions and a simplex method for a class of continuous linear programs, *Proc. 9th IFIP Conference on Optimization Techniques*, Springer-Verlag, Berlin.

ANDERSON, E. J., NASH, P., and PEROLD, A. F. (1983). Some properties of a class of continuous linear programs, *SIAM J. Control & Opt.*, **21**, 258–65.

ANDERSON, E. J., NASH, P., and PHILPOTT, A. B. (1982). A class of continuous network flow problems, *Mathematics of O. R.*, **7**, 501–14.

ANDERSON, E. J., and PHILPOTT, A. B. (1984). Duality and an algorithm for a class of continuous transportation problems, *Mathematics of O. R.*, **9**, 222–31.

ANDERSON, E. J., and PHILPOTT, A. B. (1986). *Continuous Network Programming*, Technical Report F-CAMS/TR259, Cambridge University Engineering Dept.

ANSTREICHER, K. M. (1983). Generation of feasible descent directions in continuous-time linear programming, Tech. Report SOL83-18, Dept. Operations Research, Stanford University.

APOSTOL, T. M. (1957). *Mathematical Analysis*, Addison-Wesley, Reading, Mass.

APPEL, P. (1887). Mémoire sur les deblais et les remblais, Mémoires a l'Académie des Sciences, **29**.

APPEL, P. (1928). Le problème géométrique des deblais et remblais, *Memorial des Sciences Mathématiques*.

AUBIN, J. P. (1971). A pareto minimum principle, in H. W. Kuhn and G. P. Szego (eds.), *Differential Games and Related Topics*, North-Holland, Amsterdam.

BELLMAN, R. (1957). *Dynamic Programming*, Princeton University Press, Princeton, N. J.

BEN-ISRAEL, A., and CHARNES, A. (1968). On the intersection of cones and subspaces, *Bull. Amer. Math. Soc.*, **74**, 541–44.

BEN-ISRAEL, A., CHARNES, A., and KORTANEK, K. O. (1969/70). Duality and asymptotic solvability over cones, *Bull. Amer. Math. Soc.*, **75**, 318–24. & *Erratum*, **76**, 426.

BEN-ISRAEL, A., CHARNES, A., and KORTANEK, K. O. (1971). Asymptotic duality over closed convex sets, *J. Math. Anal. Appl.*, **35**, 677–91.

BODO, E. P., and HANSON, M. A. (1978). A class of continuous non-linear complementary problems, *J. Opt. Th. Appl.*, **24**, 243–61.

BOIKOV, V. A., and CHARNYI, V. I. (1979). Decomposition of linear functional problems II: linear dynamic problems, *Aut. & Rem. Control*, **39**, 1189–96.

BORWEIN, J. M. (1981). Direct theorems in semi-infinite convex programming, *Math. Programming*, **21**, 301–18.

BORWEIN, J. M. (1983a). Semi- infinite programming duality: How special is it?, in A. V. Fiacco and K. O. Kortanek (eds.), *Semi-infinite Programming and Applications*, Springer-Verlag, Berlin.

BORWEIN, J. M. (1983b). Adjoint process duality, *Mathematics of O.R.*, **8**, 403–34.

BORWEIN, J. M., and KORTANEK, K. O. (1983). Fenchel duality and separably infinite programs, *Math Op. Statist. Ser. Opt.*, **14**, 37–48.

BRAUNSCHWEIGER, C. C., and CLARK, H. E. (1962). An extension of the Farkas theorem, *Amer. Math. Monthly*, **69**, 272–77.

BURKARD, R. E., HAMACHER, H., and TIND, J. (1982). On abstract duality in mathematical programming, *Z. für Operations Res.*, **26**, 197–209.

CHARNES, A., COOPER, W. W., and KORTANEK, K. O. (1963). Duality in semi-infinite programs and some works of Haar and Caratheodory, *Management Sci.*, **9**, 209–28.

CHARNES, A., COOPER, W. W., and KORTANEK, K. O. (1965). On representations of semi-infinite programs which have no duality gaps, *Management Sci.*, **12**, 113–21.

CHARNES, A., GRIBIK, P. R., and KORTANEK, K. O. (1980). Separably infinite programs, *Z. Operations Res.*, **24**, 33–45.

CHARNYI, V. I. (1975). Certain numerical methods for constructing an optimal economic development plan III, *Aut. & Rem. Control*, **36**, 1643–55.

CHARNYI, V. I. (1976). Functional methods of solution of linear dynamic problems of economic planning, *Aut. & Rem. Control*, **37**, 1895–1905.

CHARNYI ,V. I. (1978). Decomposition of linear functional problems I: linear programming in Hilbert space, *Aut. & Rem. Control*, **39**, 874–82.

CHOQUET, G. (1953). Theory of capacities, *Ann. Inst. Fourier Grenoble*, **5**, 131–295.

CHRISTIANSEN, E. (1980). Limit analysis for plastic plates, *SIAM J. Math. Anal.*, **11**, 514–22.

CHUKANOV, S. V. (1976). Examples of optimal trajectories in continuous-time input-output models, *Aut. & Rem. Control*, **37**, 69–78.

CHUKANOV, S. V. (1978a). On the continuous dependence of solutions of the optimal economic planning problem on the problem conditions, *Aut. & Rem. Control*, **39**, 714–23.

CHUKANOV, S. V. (1978b). Continuous approximations of solutions of linear dynamic problems of economic planning, *Aut. & Rem. Control*, **39**, 401–5.

COOPE, I. D., and WATSON, G. A. (1985). A projected Lagrangian algorithm for semi-infinite programming, *Math. Programming*, **32**, 337–56.

CORLEY, H. W., and ROBERTS, S. D. (1972). A partitioning problem with applications in regional design, *Operations Res.*, **20**, 1010–19.

CRAVEN, B. D. (1972). Nonlinear programming in locally convex spaces, *J. Opt. Th. Appl.*, **10**, 197–210.

DANTZIG, G. (1963). *Linear Programming and Extensions*, Princeton University Press, Princeton, N. J.

DREWS, W. P. (1974). A simplex like algorithm for continuous time linear optimal control problems, in R. W. Cottle and J. Kramp (eds.), *Optimization Methods in Resource Allocation*, Crane Russak, New York.

DUBOVSKII, S. V., DYUKALOV, A. N., IVANOV, Yu. N., TOKAREV, V. V., UZDEMIR, A. P., and FATKIN, Yu. M. (1973). Construction of an optimal economic plan, *Aut. & Rem. Control*, **34**, 1336–49.

DUFFIN, R. J. (1956). Infinite programs, in H. W. Kuhn and A. W. Tucker (eds.), *Linear Inequalities and Related Systems*, Princeton Univ. Press, Princeton, N. J.

DUFFIN, R. J. (1967). An orthogonality theorem of Dines related to moment problems and linear programming, *J. Combin. Th.*, **2**, 1–26.

DUFFIN, R. J., JEROSLOW, R. G., and KARLOVITZ, L. A. (1983). Duality in semi-infinite linear programming, in A. V. Fiacco and K. O. Kortanek (eds.), *Semi-infinite programming and applications*, Springer-Verlag, Berlin.

DUFFIN, R. J., and KARLOVITZ, L. A. (1965). An infinite linear program with a duality gap, *Management Sci.*, **12**, 122–34.

DUFFIN, R. J., and KARLOVITZ, L. A. (1968). Formulation of linear programs in analysis, *SIAM J. Appl. Math.*, **16**, 662–75.

DUPIN, C. (1818). Exercices de Géométrie et d'algèbre.

DYUKALOV, A. N. (1976). An optimality test in linear dynamic problems of economic planning, *Aut. & Rem. Control*, **37**, 1460–70.

DYUKALOV, A. N., and ILYUTOVICH, A. E. (1977). Indicator of optimality in nonlinear control problems with mixed constraints, *Aut. & Rem. Control*, **38**, 381–9.

FAN, K. Y. (1965). A generalization of an Alaoglu–Bourbaki theorem and its applications, *Math. Zeitschrift.*, **88**, 48–60.

FAN, K. Y. (1969). Asymptotic cones and duality of linear relations, *J. Approx. Theory*, **2**, 152–9.

FARKAS, J. (1902). Uber die Theorie der einfachen Ungleichungen, *J. Reine Angewandte Math.*, **124**, 1–27.

FARR, W. H., and HANSON,M. A. (1974a). Continuous time programming with nonlinear constraints, *J. Math. Anal. Appl.*, **45**, 96–115.

FARR, W. H., and HANSON, M. A. (1974b). Continuous time programming with nonlinear time-delayed constraints, *J. Math. Anal. Appl.*, **46**, 41–61.

FORD, L. R., and FULKERSON, D. R. (1962). *Flows in Networks*, Princeton University Press, Princeton, N. J.

GLASHOFF, K. (1979). Duality theory of semi-infinite programming, in R Hettich (ed.) *Semi-infinite Programming*, Springer-Verlag, Berlin.

GLASHOFF, K., and GUSTAFSON, S-A. (1983). *Linear Optimization and Approximation*, Springer-Verlag, New York.

GOMORY, R. E., HU, T. C., and YOHE, M. (1973). *R-separating Sets*, University of Wisconsin–Madison, MRC Report 939.

GRIBIK, P. R. (1979). A central cutting plane algorithm for semi-infinite programming problems, in R. Hettich (ed.), *Semi-infinite Programming*, Springer-Verlag, Berlin.

GRINOLD, R. C. (1968). *Continuous Programming*, Unpublished Ph. D. thesis, Operations Research Centre, Univ. California, Berkeley.

GRINOLD, R. C. (1969a). Continuous programming, part one: linear objectives, *J. Math. Anal. Appl.*, **28**, 32–51.

GRINOLD, R. C. (1969b). Continuous programming, part two: nonlinear objectives, *J. Math. Anal. Appl.*, **27**, 639–55.

GRINOLD, R. C. (1970). Symmetric duality for continuous linear programs, *SIAM J. Appl. Maths.*, **18**, 84–97.

GUSTAFSON, S. A. (1970). On the computational solution of a class of generalized moment problems, *SIAM J. Numer. Anal.*, **7**, 343–57.

GUSTAFSON, S. A. (1983). A three phase algorithm for semi-infinite programs, in A. V. Fiacco and K. O. Kortanek (eds.), *Semi-infinite Programming and Applications*, Springer-Verlag, Berlin.

GUSTAFSON, S. A., and KORTANEK, K. O. (1973). Numerical treatment of a class of semi-infinite programming problems. Nav. Res. Log. Q., **20**, 477–504.

HAAR, A. (1924). Uber lineare Ungleichungen, *Acta. Math.* (Szeged), **2**, 1–14.

HAJEK, B., and OGIER, R. G. (1984). Optimal dynamic routing in communication networks with continuous traffic, *Networks*, **14**, 457–87.

HALPERN, J. (1979). Generalized dynamic flows, *Networks*, **9**, 133–167.

HANSON, M. A. (1968). Duality for a class of continuous programming problems, *SIAM J. Appl. Maths.*, **16**, 318–23.

HANSON, M. A., and MOND, B. (1968). A class of continuous convex programming problems, *J. Math. Anal. Appl.*, **22**, 427–37.

HARTBERGER, R. J. (1974). Representation extended to continuous time, in R. W. Cottle and J. Kramp (eds.), *Optimization Methods in Resource Allocation*, Crane Russak, New York.

HEMP, W. S. (1973). *Optimum Structures*, Clarendon Press, Oxford, (1973).

HETTICH, R. (1983). A review of numerical methods for semi-infinite optimization, in A. V. Fiacco and K. O. Kortanek (eds.), *Semi-infinite Programming and Applications*, Springer-Verlag, Berlin.
HOFFMAN, K. H., and KLOSTERMAIR, A. (1976). A semi-infinite linear programming procedure and applications to approximation problems in optimal control, in G. G. Lorentz (ed.), *Approximation Theory II*, Academic Press, New York.
HOLMES, R. B. (1975). *Geometric Functional Analysis and its Applications*, Springer-Verlag, New York.
HU, T. C. (1969). *Integer Programming and Network Flows*, Addison-Wesley, Reading, Mass.
HURWICZ, L. (1958). Programming in Linear Spaces, in K. J. Arrow, L. Hurwicz, and H. Uzana (eds.), *Studies in Linear and Non-linear Programming*, Stanford.
ILYUTOVICH, A. E. (1976). Iterative optimization method for linear programming problems in functional space, *Aut. & Rem. Control*, **37**, 573–83.
ILYUTOVICH, A. E. (1980). Piecewise-continuous solutions of linear dynamic problems in economics planning. 1., *Aut. & Rem. Control*, **41**, 501–8.
IRI, M. (1979). Theory of flows in continua as approximation to flows in networks, in A. Prekopa (ed.), *Survey of Mathematical Programming*, North Holland.
ISII, K. (1960). The extrema of probability determined by generalised moments (1): bounded random variables, *Ann. Inst. Statist. Math. Tokyo*, **12**, 119–33.
JACOBS, K. (1979). A continuous max-flow problem, in O. Moeschlin and D. Pallaschke (eds.), *Game Theory and Related Topics*, North-Holland.
JACOBS, K., and SEIFFERT, G. (1983). A measure-theoretical max-flow problem, *Bull. Inst. Math. Acad. Sinica*, **11**, 261–80.
KALLINA, C., and WILLIAMS, A. C. (1971). Linear programming in reflexive spaces, *SIAM Review*, **13**, 350–76.
KANTOROVICH, L. V. (1942). On the translocation of masses, *Dokl. Akad. Nauk. SSSR*, **37**, 199–201.
KANTOROVICH, L. V., and RUBINSTEIN, G. S. (1958). On a space of completely additive functions, *Vestnik Leningrad Univ. Math.*, **13**, 52–9 (Russian).
KARNEY, D. F. (1981). Duality gaps in semi-infinite linear programming- an approximation problem, *Math. Programming*, **20**, 129–43.
KARR, A. (1983). Extreme points of certain sets of probability measures, *Mathematics of O.R.*, **8**, 74–85.
KEMPERMAN, J. H. B. (1983). On the role of duality in the theory of moments, in A. V. Fiacco and K. O. Kortanek (eds.), *Semi-infinite Programming and Applications*, Springer-Verlag, Berlin.
KOMIYA, H. (1981). Closedness of cones, *J. Math. Anal. Appl.*, **81**, 320–26.
KORTANEK, K. O. (1974). Classifying convex extremum problems over linear topologies having separation properties, *J. Math. Anal. Appl.*, **46**, 725–55.
KORTANEK, K. O. (1977). Constructing a perfect duality in infinite programming, *Applied Math. & Opt.*, **3**, 357–72.
KORTANEK, K. O. (1985). Semi-infinite programming and continuum mechanics, in E. J. Anderson and A. B. Philpott, (eds.), *Infinite Programming*, Springer-Verlag, Berlin.
KORTANEK, K. O., and YAMASAKI, M. (1980). Equalities in transportation problems and characterizations of optimal solutions, *Nav. Res. Log. Quart.*, **27**, 589–95.
KORTANEK, K. O., and YAMASAKI, M. (1982). Semi-infinite transportation problems, *J. Math. Anal. Appl.*, **88**, 555–65.
KRABS, W. (1979). *Optimization and Approximation*, Wiley, New York.
KRETSCHMER, K. S. (1961). Programmes in paired spaces, *Canadian J. Math.*, **13**, 221–38.
LANG, S. (1983) *Real Analysis*. Addison-Wesley, London.
LARSEN, A., and POLAK, E. (1966). Some sufficient conditions for continuous linear programming problems, *Int. J. Eng. Science*, **4**, 583–604.

LEHMAN, R. S. (1954). *On the Continuous Simplex Method*, RAND Research Memorandum RM1386, Santa Monica, California.

LEVIN, V. L. (1975). On the mass transfer problem *Soviet Math. Dokl.*, **16**, 1349–53.

LEVIN, V. L., and MILYUTIN, A. A. (1978). The problem of mass transfer with a discontinuous cost function and a mass statement of the duality problem for convex extremal problems, *Russ. Math. Surveys*, **34**, 1–78.

LEVINE, P., and POMEROL, J.-Ch. (1974). Infinite programming and duality in topological vector spaces, *J. Math. Anal. Appl.*, **46**, 75–89.

LEVINE, P., and POMEROL, J.-Ch. (1981). Sufficient conditions for Kuhn–Tucker vectors in convex programming, *SIAM J. Contr. & Opt.*, **17**, (1979) 689–99. & Erratum, *SIAM J. Contr. & Opt.*, **19**, 431–2.

LEVINSON, N. (1966a). A class of continuous linear programming problems, *J. Math. Anal. Appl.*, **16**, 73–83.

LEVINSON, N. (1966b). Linear programming in complex space, *J. Math. Anal. Appl.*, **14**, 44–62.

LEWIS, A. S. (1985). Extreme points of infinite transportation problems, *10th Symposium on Operations Research*, Univ. Munich.

LOWE, T. J., and HURTER, A. P. (1976). The generalized market area problem, *Management Sci.*, **22**, 1105–15.

LUENBERGER, D. G. (1969). *Optimization by Vector Space Methods*, Wiley, New York.

LUENBERGER, D. G. (1973). *Introduction to Linear and Non-linear Programming*, Addison-Wesley, Reading MA, (1973).

MEEROV, M. V., BERSHCHANSKIY, Ya. M., and LITVAK, M. L. (1980). Optimization methods for one class of linear multivariable systems, *Int. J. Control*, **31**, 239–70.

MEIDAN, R., and PEROLD, A. F. (1983). Optimality conditions and strong duality in abstract and continuous time linear programming, *J. Opt. Th. Appl.*, **40**, 61–75.

MICHELL, A. G. M. (1904). The limits of economy of material in frame-structures, *Phil. Mag.*, **8**, 589–97.

MOKHOV, V. N. (1977). Optimality criterion for an economic planning problem, *Aut. & Rem. Control*, **38**, 1018–21.

MONGE, G. (1781). *Mémoires de l'Académie des Sciences.*

MURAKAMI, A., and YAMASAKI, M. (1969). Duality theorems for continuous linear programming problems, *J. Sci. Hiroshima Univ.* Ser. A-1, **33**, 213–21.

NAKAMURA, T., and YAMASAKI, M. (1979). Sufficient conditions for duality theorems in infinite linear programming problems, *Hiroshima Math. J.*, **9**, 323–34.

NOZAWA, R. (1981). Programmings with constraints of convex processes, *Hiroshima Math. J.*, **11**, 369–78.

OGIER, R. G. (1985). *Flow Optimization in Dynamic and Continuous Networks*, Ph. D. thesis, Univ. Illinois at Urbana-champaign.

OHTSUKA, M. (1966a) Generalized capacity and duality theorem in linear programming, *J. Sci. Hiroshima Univ.* Ser. A-1, **30**, 45–56.

OHTSUKA, M. (1966b). A generalization of duality theorem in the theory of linear programming, *J. Sci. Hiroshima Univ.* Ser. A-1, **30**, 31–9.

PAPAGEORGIOU, N. S. (1982). A class of infinite dimensional linear programming problems, *J. Math. Anal. Appl.*, **87**, 228–45.

PEROLD, A. F. (1978). *Fundamentals of a Continuous Time Simplex Method*, Tech. Report, SOL 78-26, Dept. of Operations Research, Stanford University.

PEROLD, A. F. (1981). Extreme points and basic feasible solutions in continuous time linear programming, *SIAM J. Control and Opt.*, **19**, 52–63.

PESTRYAKOV, A. K. (1977). Numerical solution to dynamic problems of economic planning by method of remote point, *Aut. & Rem. Control*, **38**, 1022–31.

PESTRYAKOV, A. K. (1978). Decomposition of linear functional control functions, *Aut. & Rem. Control*, **39**, 1504–13.

PESTRYAKOV, A. K. (1980). One method of solving linear dynamic control problems, *Aut. & Rem. Contr.*, **41**, 981–89.

PHILPOTT, A. B. (1982). Algorithms for continuous network flow problems, Unpublished Ph.D. thesis, University of Cambridge.

PHILPOTT, A. B. (1985). Network programming in continuous time with node storage, in E. J. Anderson and A. B. Philpott (eds.), *Infinite Programming*, Springer-Verlag, Berlin.

POMEROL, J. Ch. (1985). Openness, closedness and duality in Banach spaces with applications to continuous linear programming, in E. J. Anderson and A. B. Philpott (eds.), *Infinite Programming*, Springer-Verlag, Berlin.

PONSTEIN, J. (1980). *Approaches to the Theory of Optimization*, Cambridge University Press, Cambridge.

POWELL, M. J. D. (1981). *Approximation Theory and Methods*, Cambridge University Press, Cambridge.

PRAGER, W. (1974). *Introduction to Structural Optimization*, Udine Lecture Notes, Springer-Verlag, Berlin.

RACHEV, S. T. (1985). The Monge-Kantarovich mass transference problem and its stochastic applications, *SIAM J. Th. Prob. Appl.*, **29**, 647–76.

REILAND, T. W. (1980a). Optimality conditions and duality in continuous programming, I. Convex Programs and a Theorem of the Alternative, *J. Math. Anal. Appl.*, **77**, 297–325.

REILAND, T. W. (1980b). Optimality conditions and duality in continuous programming, II. The Linear Problem Revisited, *J. Math. Anal. Appl.*, **77**, 329–43.

RIESZ, M. (1938). Integrales de Riemann-Liouville et potentiels, *Acta Sci.* (Szeged), **9**, 1–42.

RITTER, K. (1969a). Optimisation theory in linear spaces, 1, *Math. Ann.*, **182**, 189–206.

RITTER, K. (1969b). Optimisation theory in linear spaces, 2, *Math. Ann.*, **183**, 169–80.

ROBERTSON, A. P., and ROBERTSON, W. J. (1973). *Topological Vector Spaces*, Cambridge University Press, Cambridge.

ROCKAFELLAR, R. T. (1967). Duality and stability in extremum problems involving convex functions, *Pacific J. Math.*, **21**, 167–87.

ROCKAFELLAR, R. T. (1970). *Convex Analysis*, Princeton University Press, Princeton, N.J.

ROGOSINSKI, W. W. (1958). Moments of non-negative mass, *Proc. Roy. Soc.* (London) Ser. A., **245**, 1–27.

ST, GERMAINE, A. DE,. (1886). Etude sur le problème des deblais et remblais *Mémoires de l'Académie Nationale des Sciences Arts et Belles Letters de Caen.*

SCHAEFER, H. H. (1971). *Topological Vector Spaces*, Springer-Verlag, New York.

SCHECHTER, M. (1972a). Duality in continuous linear programming, *J. Math. Anal. Appl.*, **37**, 130–41.

SCHECHTER, M. (1972b). Linear programs in topological vector spaces, *J. Math. Anal. Appl.*, **37**, 492–500.

SCHIROTZEK, W. (1981). On Farkas type theorems, *Comment. Math. Univ. Carolin.*,**22**, 1–14.

SEGALL, A. (1977). The modelling of adaptive routing in data communicating networks, *IEEE Trans. Comm.*, **25**, 85–95.

SEGERS, R. G. (1974). A generalised function setting for dynamic optimal control problems, in R. W. Cottle and J. Kramp (eds.), *Optimisation methods in Resource Allocation*, Crane Russak, New York.

SETHI, S. P., DREWS, W. P., and SEGERS, R. G. (1982). A unified framework for linear control problems with state-variable inequality constraints *J. Opt. Th. Appl.*, **36**, 93–109.

SINGH, C. (1978). A sufficient optimality criteria in continuous time programming for generalised convex functions, *J. Math. Anal. Appl.*, **62**, 506–11.

SINGH, C., and FARR, W. H. (1977). Saddlepoint optimality criteria of continuous time programming without differentiability, *J. Math. Anal. Appl.*,**59**, 442–53.

STIEFEL, E. (1960). Note on Jordan elimination, linear programming and Tchebycheff approximation, *Numerische Math.*, **2**, 1–17.
STRANG, G. (1983). Maximal flow through a domain, *Math. Programming*, **26**, 123–43.
STRANG, G., and KOHN, R. V. (1983). Hencky-Prandtl nets and constrained Mitchell trusses, *Comp. Methods Appl. Mech. Eng.*, **36**, 207–22.
TEMAM, R. (1983). Dual variational principles in mechanics and physics, in A. V. Fiacco and K. O. Kortanek (eds.), *Semi-infinite Programming*, Springer-Verlag, Berlin.
TEMAM, R., and STRANG, G. (1980). Duality and relaxation in the variational problems of plasticity, *J. de Mechanique*, **19**, 493–527.
TODD, M. J. (1978). Solving the generalized market area problem, *Management Sci.*, **24**, 1549–54.
TROLTZSCH, F. (1980). Duality theorems for a class of continuous linear programming problems in a space of Bochner integrable abstract functions, *Math. Op. Statist. Ser. Optimisation*, **11**, 375–88.
TYNDALL, W. F. (1965). A duality theorem for a class of continuous linear programming problems, *SIAM J. Appl. Math.*, **13**, 644–66.
TYNDALL, W. F. (1967). An extended duality theory for continuous linear programming problems, *SIAM J. Appl. Math.*, **15**, 1294–98.
UNDERWOOD, R. G. (1976). Continuous programs and their relation to continuous games, *J. Math. Anal. Appl.*, **56**, 102–12.
VAN, SLYKE, R. M., and WETS, R. J. B. (1968). A duality theory for abstract mathematical programs with applications to optimal control theory, *J. Math. Anal. Appl.*, **22**, 679–706.
VOLKOV, YU. N. (1980). Investigation of a linear dynamic problem in economic planning, *Aut. & Rem. Contr.*, **41**, 967–73.
WATSON, G. A. (1975). A multiple exchange algorithm for multivariate Chebyshev approximation, *SIAM J. Numer. Anal.*, **12**, 46–52.
WATSON, G. A. (1985). Lagrangian methods for semi-infinite programming problems, in E. J. Anderson and A. B. Philpott (eds.), *Infinite Programming*, Springer-Verlag, Berlin.
WHITTLE, P. (1971). *Optimization Under Constraints*, Wiley, London.
WU, S. Y. (1985). The general capacity problem, in P. BRUCKER (ed.), *Proc. 9th Symposium on Operations Research.*
YAMASAKI, M. (1966). On a capacitability problem raised in connection with linear programming, *J. Sci. Hiroshima Univ.* Ser. A-1, **30**, 57–73.
YAMASAKI, M. (1968). Duality theorems in mathematical programmings and their applications, *J. Sci. Hiroshima Univ.* Ser. A-1, **32**, 331–56.
ZALINESCU, C. (1978). A generalisation of the Farkas lemma and applications to convex programming, *J. Math. Anal. Appl.*, **66**, 651–78.
ZIMMERMANN, U. (1981). Linear and combinatorial optimization in ordered algebraic structures, *Annals Discrete Maths.*, **10**, North-Holland, Amsterdam.
ZOWE, J. (1975). A duality theorem for a convex programming problem in order complete vector lattices, *J. Math. Anal. Appl.*, **50**, 273–87.

Index of Problem Statements

Bracketed page numbers give references to the corresponding dual problems

Subject Index